MUDVILLE
BUGVILLE
PODUNK

PERPLEXING PUZZLES *and* TANTALIZING TEASERS

by Martin Gardner

ILLUSTRATED BY *Laszlo Kubinyi*

TWO VOLUMES BOUND AS ONE

DOVER PUBLICATIONS, INC., *New York*

Published in Canada by General Publishing Company, Ltd., 30 Lesmill Road, Don Mills, Toronto, Ontario.
Published in the United Kingdom by Constable and Company, Ltd.

This Dover edition, first published in 1988, is an unabridged republication in one volume of *Perplexing Puzzles and Tantalizing Teasers*, originally published by Simon & Schuster, Inc., New York, in 1969, and *More Perplexing Puzzles and Tantalizing Teasers*, originally published by Pocket Books, a division of Simon & Schuster, Inc., under their Archway imprint, in 1977.

Manufactured in the United States of America
Dover Publications, Inc., 31 East 2nd Street, Mineola, N.Y. 11501

Library of Congress Cataloging-in-Publication Data

Gardner, Martin, 1914–
Perplexing puzzles and tantalizing teasers / Martin Gardner : illustrated by Laszlo Kubinyi.
p. cm.
Reprint (1st work). Originally published: New York : Simon & Schuster, 1969.
Reprint (2nd work). Originally published: New York : Archway, 1977.
Contents: Perplexing puzzles and tantalizing teasers—More perplexing puzzles and tantalizing teasers.
Summary: Combines two previously published works, resulting in ninety-three brain-teasing puzzles with an emphasis on humor.
ISBN 0-486-25637-5 (pbk.)
1. Puzzles—Juvenile literature. [1. Puzzles.] I. Kubinyi, Laszlo, 1937– ill. II. Gardner, Martin, 1914– More perplexing puzzles and tantalizing teasers. 1988. III. Title.
GV1493.G34 1988
793.73—dc19 87-34584
CIP
AC

PERPLEXING PUZZLES
and
TANTALIZING TEASERS

FOR MY NEPHEW

Harold Berg Gardner

CONTENTS

INTRODUCTION

This is a collection of many different kinds of puzzles, some old, some new, but most of them will (I hope) be puzzles you haven't seen or heard before. None of them is very difficult, but some of the puzzles are tricky, with answers that will surprise and amuse you.

The solution to each puzzle is given at the end of the book, but of course you won't get much fun out of this collection if you start peeking at the answers before you've done your best to solve the puzzles first.

Happy puzzling!

MARTIN GARDNER

1 *Ridiculous Riddles*

1. What is green and flies through the air?
2. What is yellow and always points north?
3. What did the five-hundred-pound mouse say to the cat?
4. What is black and white and red all over?
5. Who was the tallest President of the United States?
6. How do you make a hippopotamus float?
7. How does an elephant put his trunk in the alligator's mouth?
8. What has a hump, is brown, and lives at the North Pole?
9. What's red, then purple, then red, then purple . . . ?
10. What does *decor* mean?
11. What do you sit on, sleep on, and brush your teeth with?
12. What has 2,754 seeds and moves by itself?
13. What has 18 legs, is covered with red spots, and catches flies?
14. Why do baby ducks walk softly? 
15. What is round and green, is covered with blue hair, has big scaly claws, weighs five thousand pounds, and goes peckety-peck-peck?
16. What has four stander-uppers, four puller-downers, two hookers, two lookers, and a swishy-wishy?
17. How do you top a car?
18. Why does Smokey the Bear wear a forest ranger's hat?

2 *Handies*

Have you ever heard of "handies"? That was the name of a popular puzzle game everybody was playing in the 1930's. The idea was to do something silly with your hands and ask "What's this?" People tried to guess. If they couldn't guess correctly, you told them what it was. For example, the first handie shown here is "Indian peeking over his indoor television antenna." Now see how good you are at guessing the others.

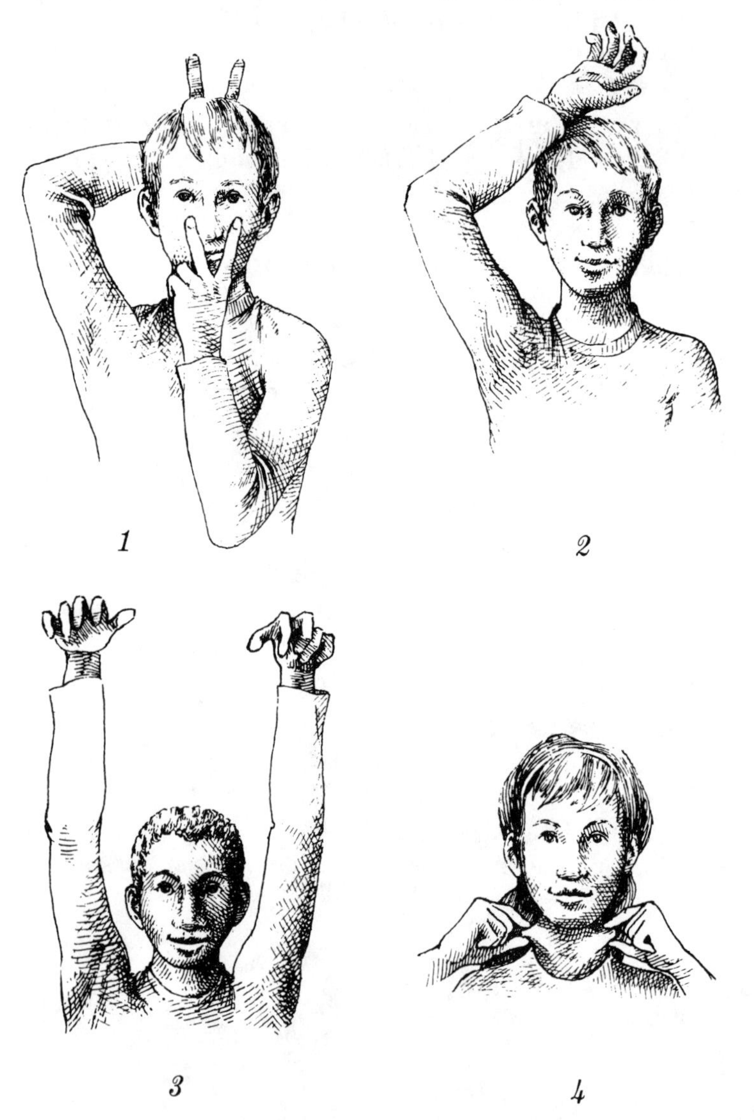

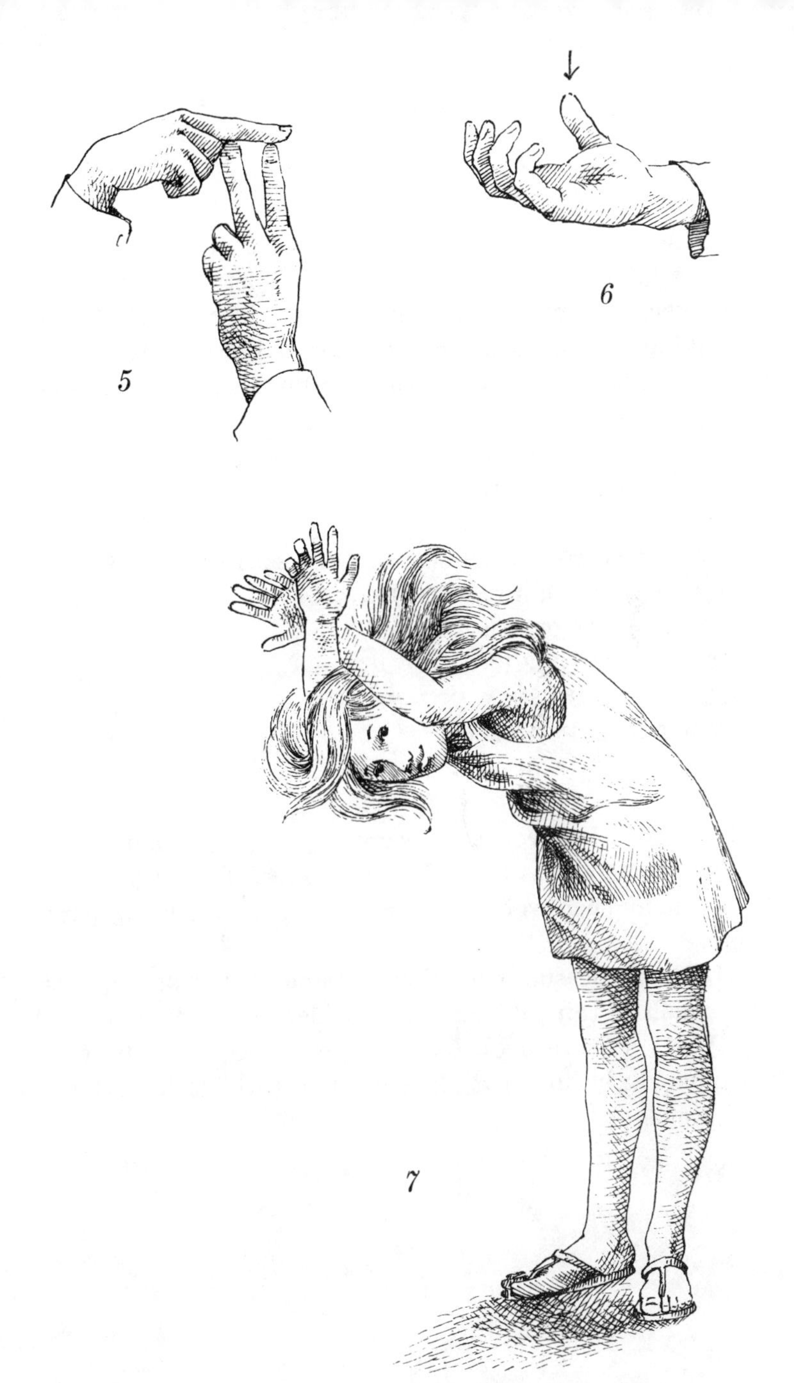
5
6
7

3 *Fun with Palindromes*

Palindromes are words or sentences that are the same when read backward or forward. The picture shows how Adam, when he first met Eve, might have introduced himself by speaking a palindrome, and how Eve might have replied by speaking another. Even the serpent is uttering a palindrome!

There are hundreds of longer sentences that read the same both ways. Here are a few clever ones:

STRAW? NO TOO STUPID A FAD. I PUT SOOT ON WARTS.
A MAN, A PLAN, A CANAL—PANAMA!
WAS IT A BAR OR A BAT I SAW?
DRAW PUPIL'S LIP UPWARD
TEN ANIMALS I SLAM IN A NET
POOR DAN IS IN A DROOP
NO, IT IS OPEN ON ONE POSITION

How good are you at recognizing a palindromic word when you come across one in your reading? To test yourself, see how many such words you can find in the following paragraphs:

"Look at the sun, over there behind that radar tower," said Hannah. "I think it looks much redder than it did at noon."

"Wow! It sure does, Ma'am," exclaimed Otto, bending over to pat the head of a small brown pup with black markings over one eye.

Madam, in Eden, I'm Adam.
Tut tut!
Eve.

4 *The Lost Star*

A perfect five-pointed star is hidden somewhere in the pattern of this patchwork quilt. Can you find it?

5 *Find the Hidden Animals*

In each of the sentences below, the name of an animal is concealed. The first sentence is marked so you can see how the word "dog" is hidden.

Can you find the animal in each of the other sentences?

1. What shall I do, Gertrude?
2. Asking nutty questions can be most annoying.
3. A gold key is not a common key.
4. Horace tries in school to be a very good boy.
5. People who drive too fast are likely to be arrested.
6. Did I ever tell you, Bill, I once found a dollar?
7. John came late to his arithmetic class.
8. I enjoy listening to music at night.

6 *Tricky Questions*

Each puzzle on this page has a funny "catch" to it. Think hard and try to guess the joke before you peek at any of the answers.

HIGGS'S PIGS

Farmer Higgs owns three pink pigs, four brown pigs, and one black pig. How many of Higgs's pigs can say that it is the same color as another pig on Higgs's farm?

PENNIES FOR SALE

Why are 1966 pennies worth almost twenty dollars?

POP AND GRANDPOP

Tom says his grandfather is only six years older than his father. Is that possible?

THROUGH THE PIPE

Jim and Tom find a long piece of pipe in a vacant lot. It's big enough so that each boy can just manage to squeeze into it and crawl from one end to the other. If Jim and Tom go into the pipe from opposite ends, is it possible for each boy to crawl the *entire length* of the pipe and come out the other end?

7 *The Five Airy Creatures*

Jonathan Swift, who wrote *Gulliver's Travels,* also wrote this clever puzzle-poem:

We are little airy creatures,
All of different voice and features;
One of us in "glass" is set,
One of us you'll find in "jet,"
T'other you may see in "tin,"
And the fourth a "box" within.
If the fifth you should pursue,
It can never fly from "you."

Can you guess who or what the five "little airy creatures" are?

8 *The Maze of the Minotaur*

An ancient Greek myth tells how Theseus found his way through a huge labyrinth, a confusing network of passageways some of which lead only to a dead end, and killed the Minotaur —a ferocious creature, half man and half bull—who lived at the center. Here is a picture showing what the plan of the labyrinth could have been. No one has ever drawn a maze that *looks* easier to work, but actually is so difficult.

Use the point of a toothpick, so you won't mark up the page and ruin the puzzle for someone else. You'll be lucky if you can find your way to the Minotaur in less than twenty minutes!

START HERE

9 *The Dime-and-Penny Switcheroo*

Put two dimes and two pennies in the spaces that contain their pictures. The object is to make the pennies and dimes change places in *exactly eight moves.*

You are allowed two kinds of moves:

1. You can *slide* any coin into an empty space next to it.

2. You can *jump* any coin over the coin next to it, like a jump in checkers, provided you land on an empty space.

The puzzle isn't as easy as it looks. Time yourself to see how long it takes you to switch the coins in eight moves. If you solve it in five minutes, you're a genius. Ten minutes is excellent. Twenty minutes is about average.

Remember, only eight moves are allowed. If you do it in *more* moves, you haven't solved the puzzle.

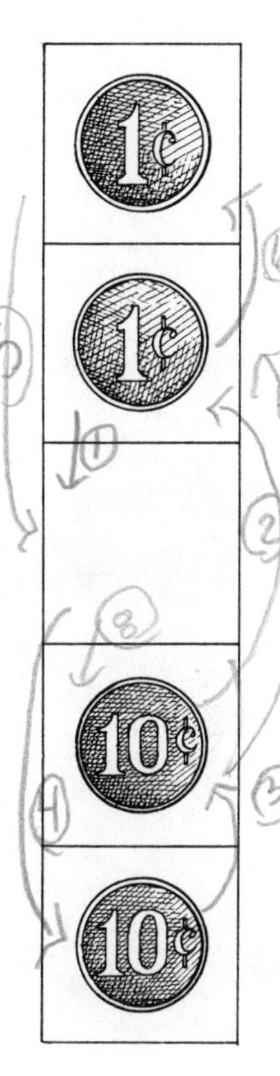

10 *A Dozen Droodles for Nimble Noodles*

"Droodles" are drawings made with just a few simple lines. The word was invented by Roger Price, who has written several very funny books about them. At first a droodle doesn't look like anything. But if you study it for a while, sometimes you will suddenly see what it is. Sometimes you'll *never* guess correctly. See how many of these droodles you can guess before you look at the answers.

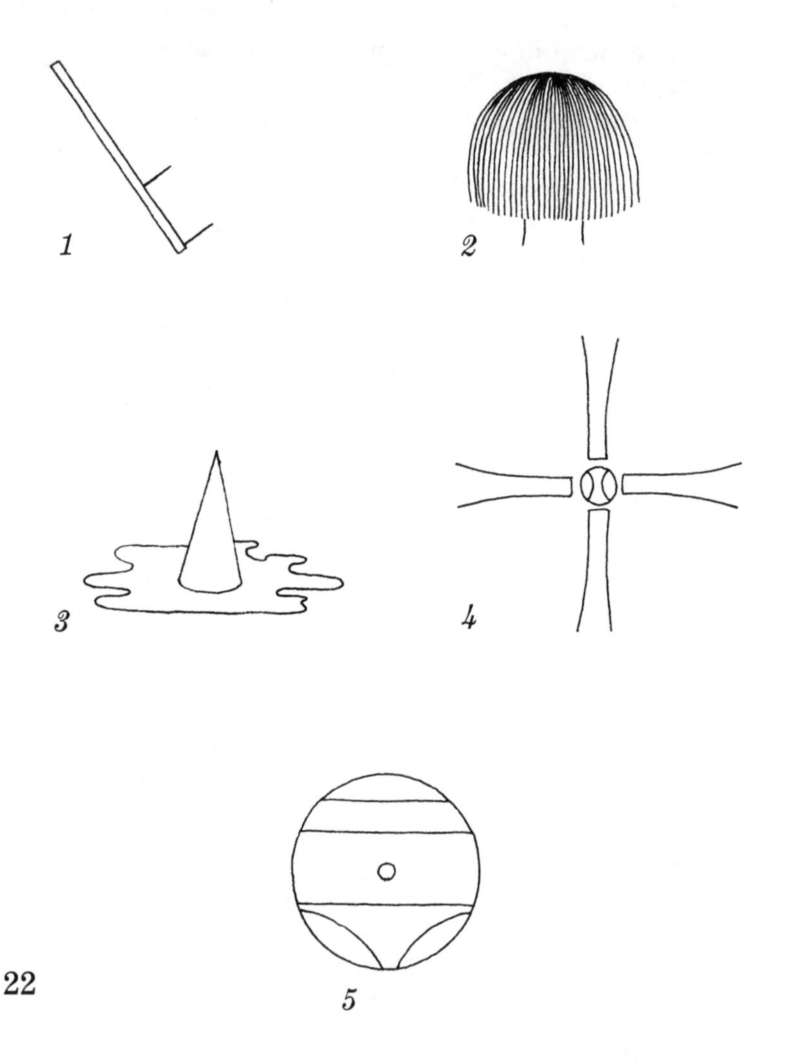

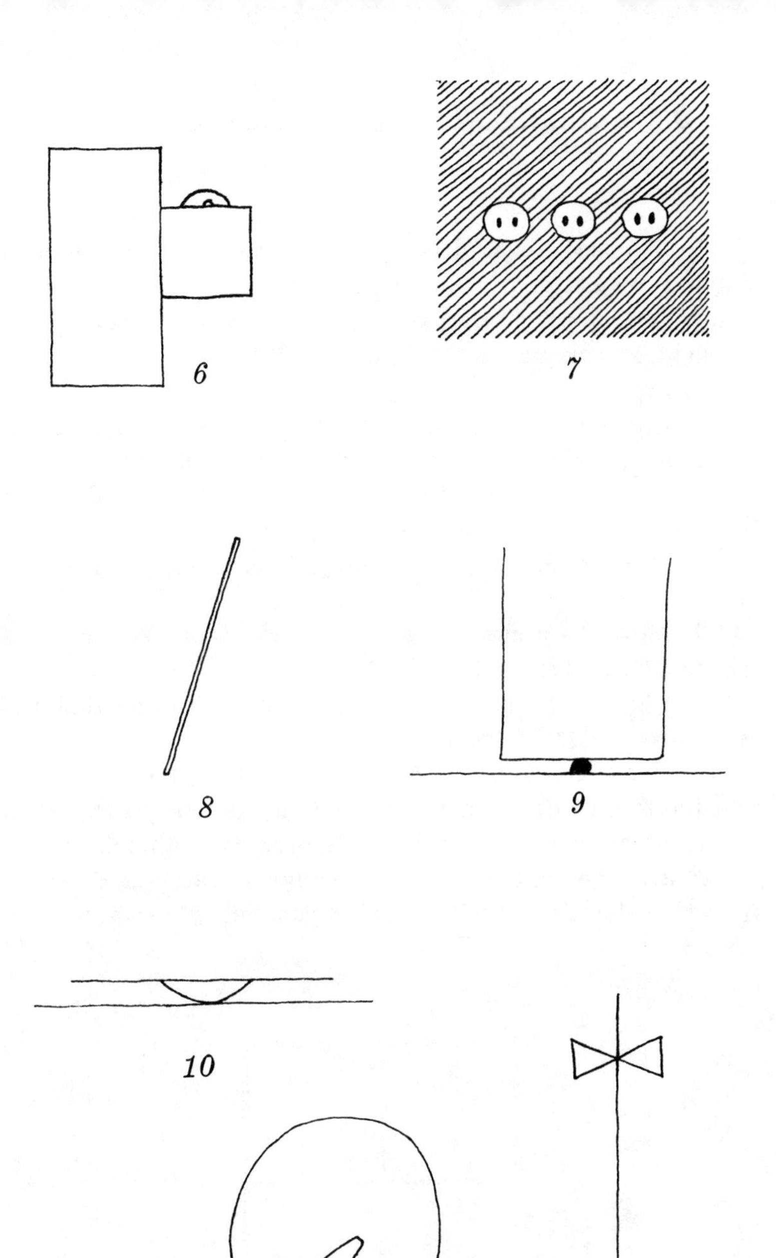
6
7
8
9
10
11
12

11 *Tantalizing Toothpick Teasers*

You'll need about fifteen toothpicks, or burned wooden matches, to test your wits on these six clever toothpick puzzles. If you can solve three, you're average. Four is good, five is excellent, and six makes you a genius.

1. Change the positions of four toothpicks to make three small squares, all the same size, and no toothpicks left over.
2. Change the positions of two toothpicks to make four small squares, all the same size, and no toothpicks left over.
3. Remove six toothpicks completely, leaving ten on the table.
4. Move the position of one toothpick and make the house face east instead of west.
5. Change the positions of three toothpicks so that the triangular pattern points down instead of up.
6. The picture shows how to make four triangles with nine toothpicks. Can you find a way to make four triangles, all the same size as the ones shown, with only six toothpicks?

Hint: The solution to this toothpick teaser is different from the other five. It will require a completely new approach.

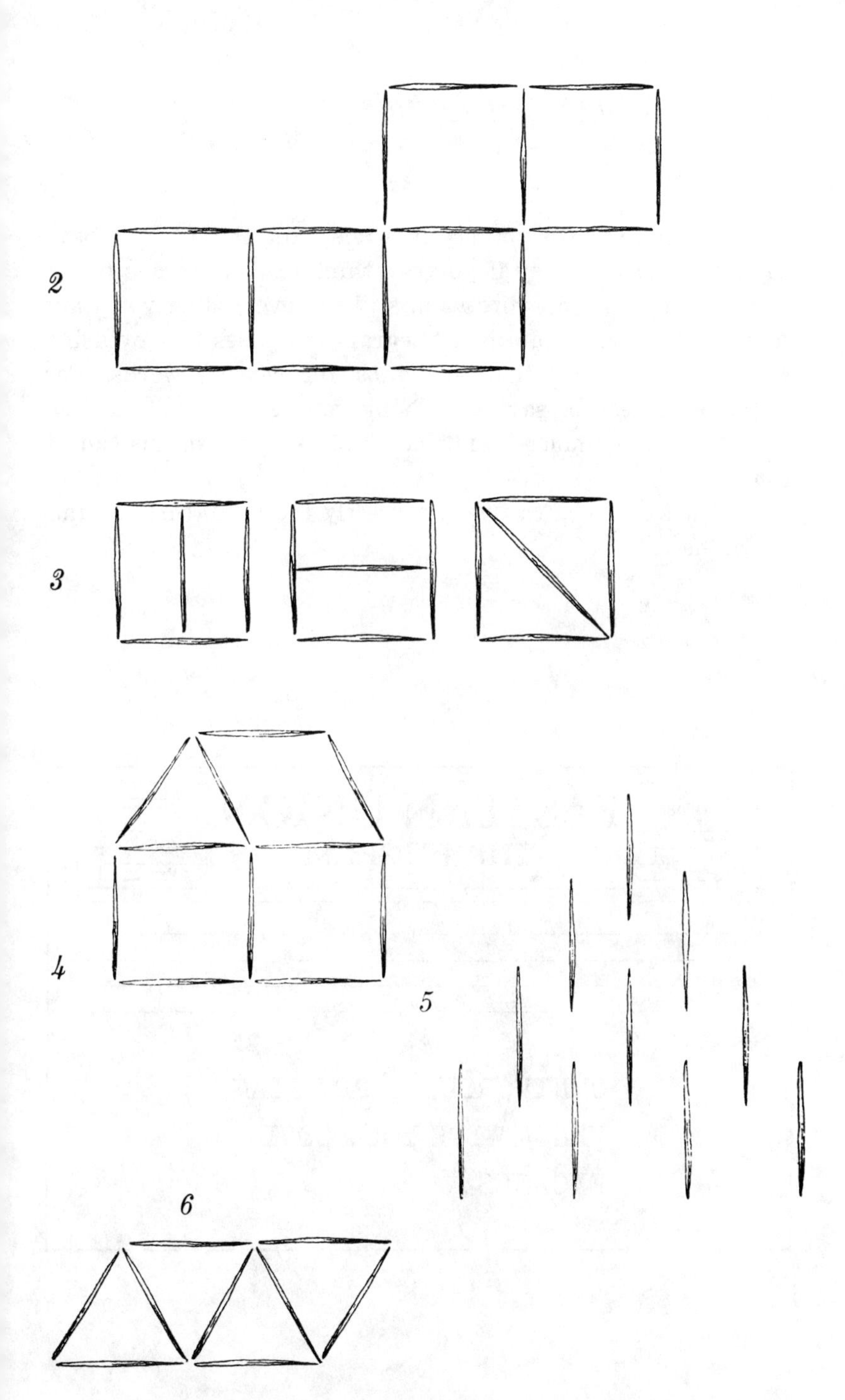
2
3
4
5
6

12 *Read the Thriftigrams*

The more words you use in a telegram, the more it costs, so you can save money if you can think of clever ways to cut down the number of words and still say everything you want to say. The thriftigram is a telegram that does this by using many single words that are puns for several words. For example, instead of saying, "Thank you very much," you can say, "Sanctuary much," and have only two words instead of four.

Now see if you can read correctly the three thriftigrams shown here.

1

DOMESTIC Check Service Desired	EASTERN UNION THRIFTIGRAM	INTERNATIONAL Check Service Desired
THRIFTIGRAM		FULL RATE
DAY LETTER		LETTER
NIGHT LETTER		SHORE-SHIP

NO. WDS.	PD. OR COLL.	NO.	CHARGE TO	TIME

*To*__________ *Date*__________

*Address*__________

*City*__________ *State*__________

MESSAGE:

OMNIVOROUS HAPPY SIAM
VENOM WITH YOU. LOVE
ENCASES.

2

DOMESTIC Check Service Desired		EASTERN UNION THRIFTIGRAM	INTERNATIONAL Check Service Desired	
THRIFTIGRAM			FULL RATE	
DAY LETTER			LETTER	
NIGHT LETTER			SHORE-SHIP	

NO. WDS.	PD. OR COLL.	NO.	CHARGE TO	TIME

*To*________________ *Date*________

*Address*________________

*City*____________ *State*________

MESSAGE:

HAVE TOOTHACHE PLANE.

CANOE MIMI AT AIRPORT?

3

DOMESTIC Check Service Desired		EASTERN UNION THRIFTIGRAM	INTERNATIONAL Check Service Desired	
THRIFTIGRAM			FULL RATE	
DAY LETTER			LETTER	
NIGHT LETTER			SHORE-SHIP	

NO. WDS.	PD. OR COLL.	NO.	CHARGE TO	TIME

*To*________________ *Date*________

*Address*________________

*City*____________ *State*________

MESSAGE:

VALUE BEMOAN VALENTINE?

OLIVE YOU.

13 *More Tricky Questions*

These are just as tricky as the "tricky questions" on earlier pages. Don't look at the answers until you have tried your best to figure out the "catch" in each one.

THE TRAMP AND THE TRAIN

A tramp was walking down a railroad track when he saw a fast express train speeding toward him. Of course, he jumped off the track. But before he jumped, he ran ten feet *toward* the train. Why?

A HARD-BOILED PROBLEM

If it takes twenty minutes to hard-boil one goose egg, how long will it take to hard-boil four goose eggs?

HEAP TOUGH PROBLEM

A big fat Indian and a small thin Indian were sitting outside a teepee, each smoking a pipe. The little Indian was the son of the big Indian, but the big Indian was *not* the little Indian's father. How come?

14 *Mr. Bushyhead's Problem*

Mr. Bushyhead was driving through a strange town when he decided to stop, park his car, and get a haircut. He asked a boy where he could find a barbershop.

"We have only two barbers in this town," said the boy. "One of them has a shop at the north end of Main Street and the other has a shop at the south end."

Mr. Bushyhead walked north on Main Street until he reached one of the barbershops. It looked as if it hadn't been cleaned in months. Cut hair was all over the floor. The barber himself needed a shave and his haircut looked terrible.

Mr. Bushyhead walked in the other direction until he came to the second barbershop. It looked neat and cheerful inside. The floor had been swept. The barber was neatly dressed, freshly shaved, and had a neat haircut.

Why did Mr. Bushyhead walk back to the *first* barbershop to get his haircut?

15 *Sneaky Arithmetic*

These are easy number problems, but if you try to answer them too quickly you'll probably make mistakes. They're fun to spring on friends.

1. How much is 1 times 2 times 3 times 4 times 5 times 6 times 7 times 8 times 9 times 0?
2. Divide 20 by $\frac{1}{2}$ and add 3. What is the result?
3. How much does a brick weigh if it weighs 5 pounds plus half its own weight?
4. A farmer had 17 sheep. All but 9 died. How many were left?
5. How much is twice one half of 987,654,321?

The Careless Sign Painter

A sign painter was in such a hurry to finish his work that he made some careless mistakes when he painted these four signs. He painted one letter of each word wrong. See if you can change the incorrect letter in each word so that all four signs read properly.

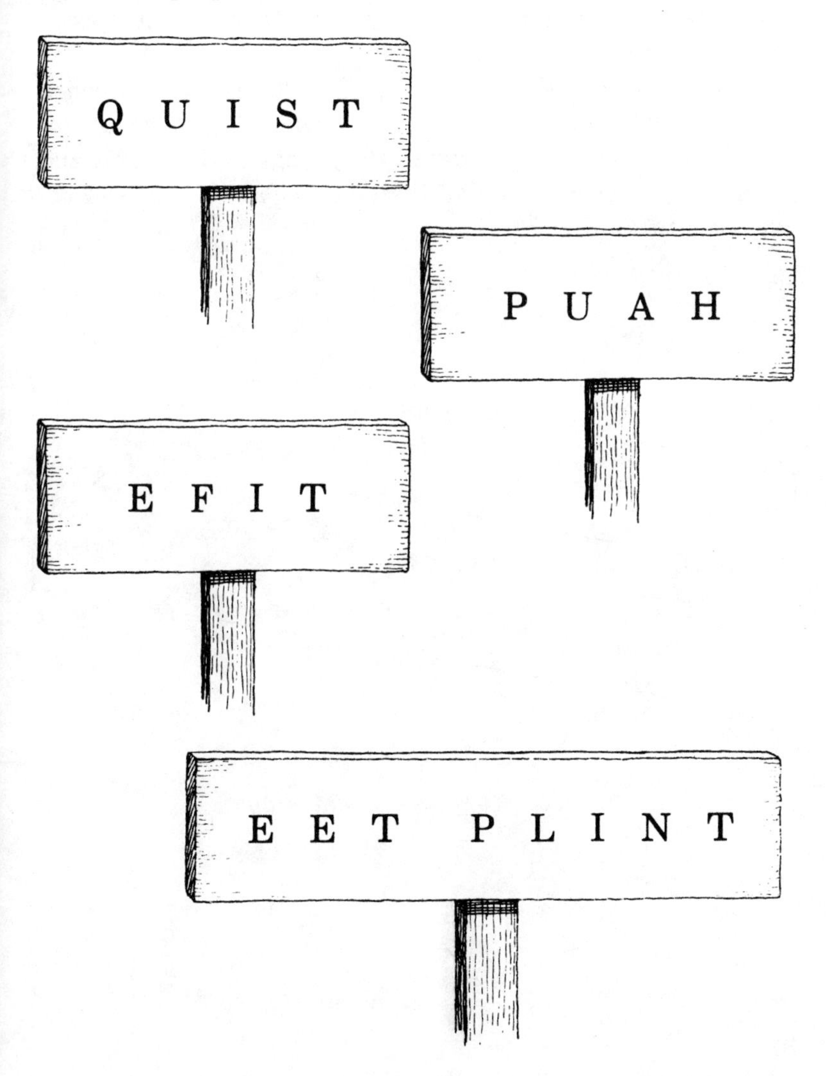

17 *The Undecidables*

Does this widget (which looks like a tool, but isn't) have two prongs or three? There isn't any answer, because if you look at one end you see three prongs; but if you look at the other end you see one slot, so there seem to be only *two* prongs, one on each end of the slot! Psychologists call it an "undecidable figure." No matter how long you study it, you just can't decide how it is constructed.

The next page shows three more undecidables: a wooden frame with three holes into which the widget's prongs are inserted, three nutty nuts to fasten the prongs to the frame, and a crazy crate for carrying all these undecidables until you decide what to do with them!

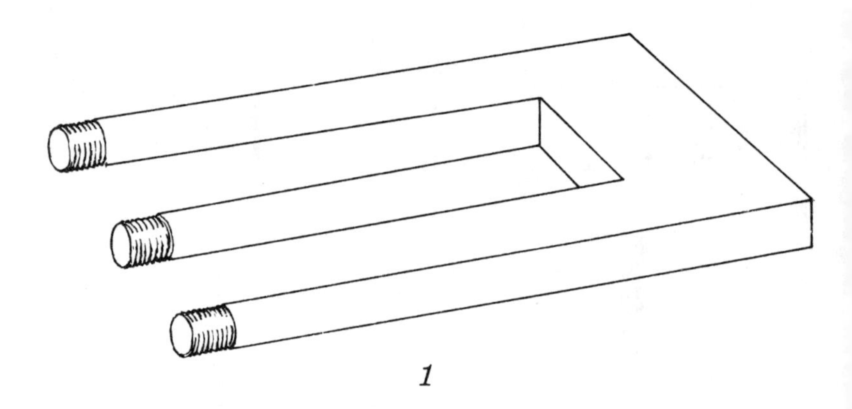

1

A three-pronged, one-slot widget.

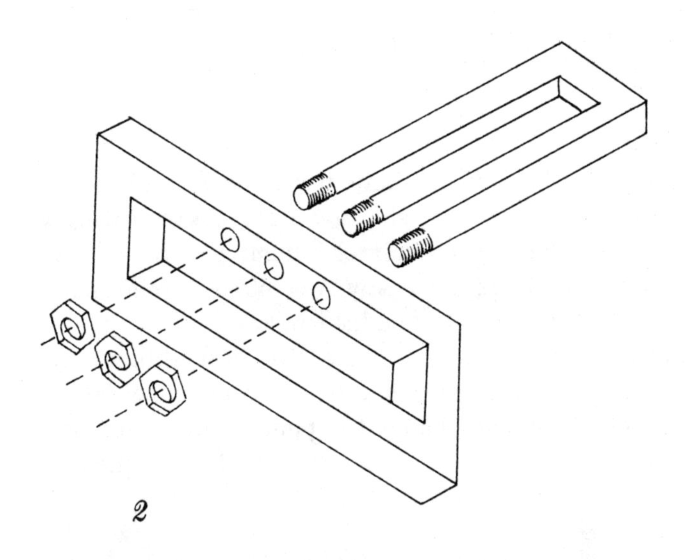

Frame for widget and nuts to hold it in place.

Crazy crate for carrying widgets and other undecidable objects.

18 *Sally's Silly Walk*

Last Sunday, when Sally went for a walk, she saw a policeman skipping rope; she saw a fire engine eating an ice-cream cone; she saw a squirrel humming a tune; she saw a puppy climbing a tree; she saw two robins playing hopscotch; she saw an organ grinder and his monkey.

Was Sally imagining all this? No, everything in the long sentence is right, except that it has been punctuated incorrectly. See if you can change the punctuation, without changing a single word, so that the sentence reads correctly.

19 *Still More Tricky Questions*

THE MISPELLED WORD

Somewhere on this page there is a word that is not spelled correctly. Can you find it?

FLAPDOODLE'S WALK

Archibald Flapdoodle walked outside through a pouring rain for twenty minutes without getting a single hair on his head wet. He didn't wear a hat, carry an umbrella, or hold anything over his head. His clothes got soaked. How could this happen?

STAMPS TO STUMP YOU

It takes twelve one-cent stamps to make a dozen. How many four-cent stamps does it take to make a dozen?

WHAT DO YOU THINK?

There once was a race horse
That won great fame.
What-do-you-think
Was the horse's name.

20 *Guess The Typitoons*

A typitoon is a picture made by hitting typewriter keys. Here are a few examples. See if you can guess what they are. If there is a typewriter in your house, perhaps you can invent some new ones.

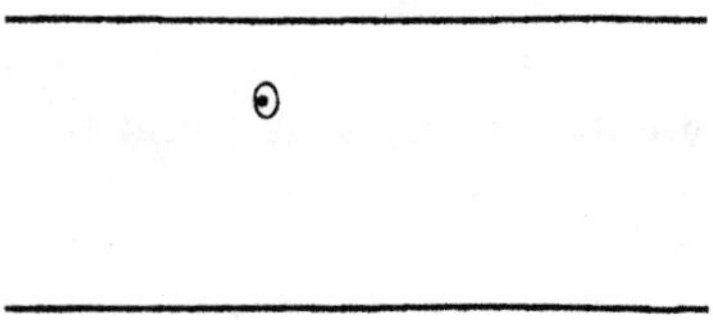

1

2

3

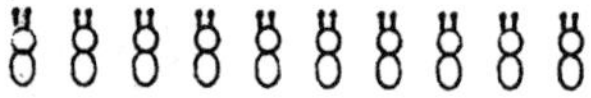

4

nnnnnnnnnhnn I

5

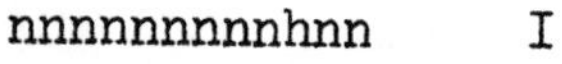

6

21 *The Fish and the Robot*

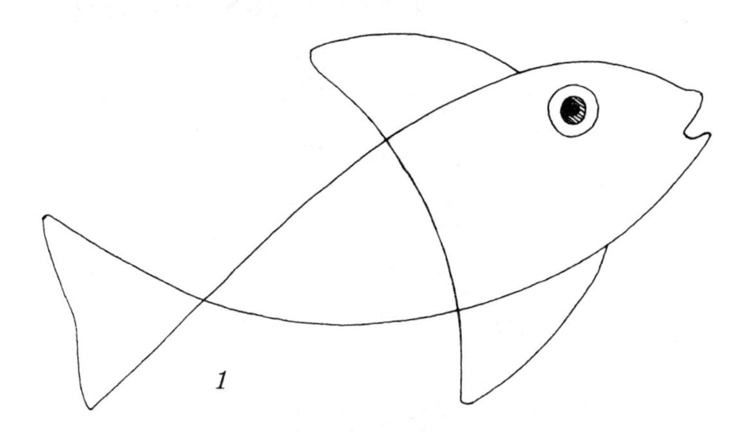
1

Get a pencil and a sheet of blank paper, then see if you can draw this fish in one continuous line without taking the pencil point off the paper, without going over any part of the line twice, and without crossing over any part of the line.

After you've learned how to draw the fish, see if you can draw the robot in the same way.

The eyes of the fish and robot are not part of either puzzle.

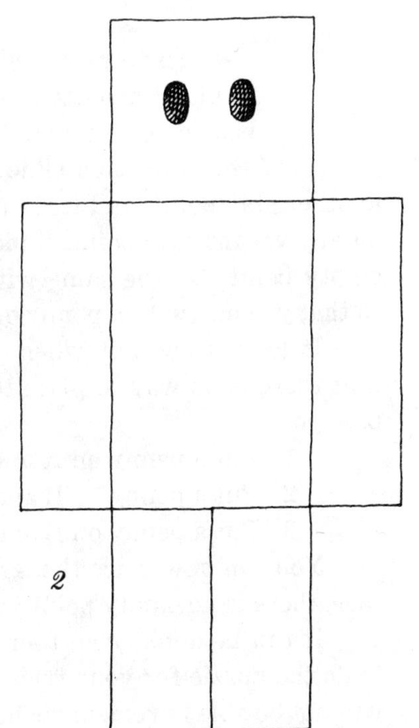
2

22

The Christmas-Star Puzzle

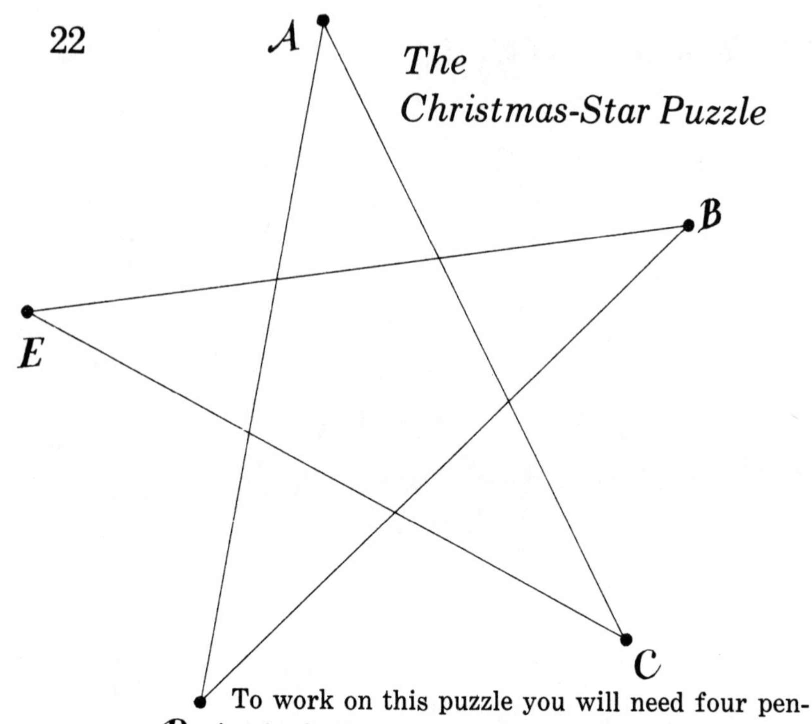

To work on this puzzle you will need four pennies (or buttons, or any four small objects you can use for counters). Put the first penny on any point of the star, then slide it along a straight black line to another star point, and leave it there. Now put a second penny on any vacant star point. Slide it along a black line to another empty point. Do the same with the third and fourth pennies, so that you end with a penny on each of four points.

It looks easy, but when you try it you are likely to find that there is no way to place the last penny. For example, suppose you:

1. Put a penny on A and slide it to C.
2. Put a penny on B and slide it to D.
3. Put a penny on B and slide it to E.

You can now place the last penny on A or B, but in either case, there is no empty point you can slide it to.

It *can* be done! And there is a secret that will enable you to do the puzzle for your friends, and do it so quickly that they will not be able to remember how you did it!

23 *The Boring Bookworm*

A ten-volume encyclopedia stands on the shelf as shown. Each volume is two inches thick. Suppose a bookworm starts at the front cover of Volume 1 and eats his way in a straight horizontal line through to the back cover of Volume 10. How far does the worm travel?

24 *The Triangular Turkey*

How many different triangles can you find in this picture of a Thanksgiving turkey?

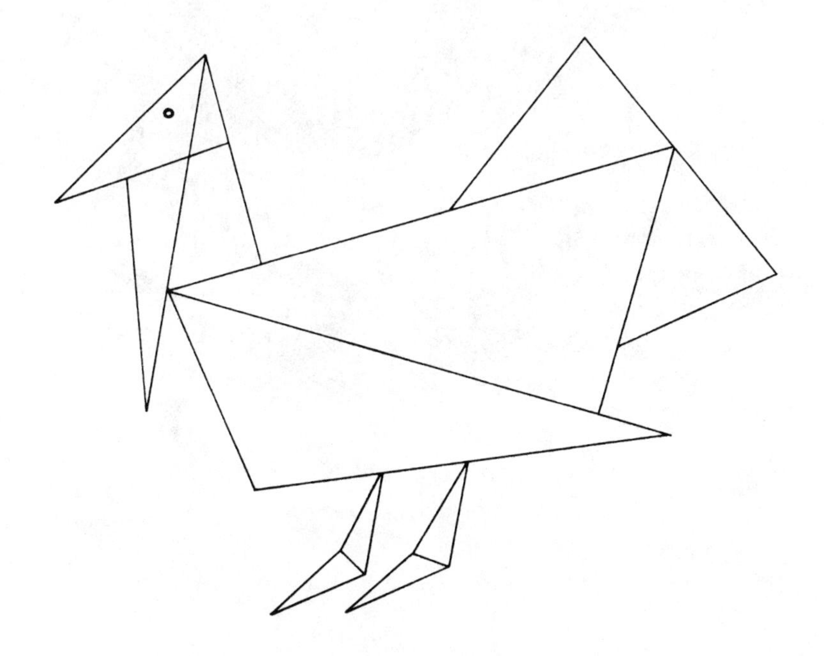

25 *Find the Best Words*

The bull in this picture has just swallowed a time bomb that is set to go off in five minutes. Which of the four words below do you think best describes the situation?

Awful
Abominable
Dreadful
Shocking

Here is a picture of the same scene, after the bomb went off. Which word below is the best description of the picture?

Amazing
Silly
Messy
Noble

Zoo-lulus were created by Max Brandel for *Mad Magazine.** What is a zoo-lulu? It is a printed name of an animal with something added or done to it that makes you think of that animal. For example, you can stretch the *H in dac*⊢——⊣*shund* so that the name *looks* like a dachshund! Or you can make the first two *o*'s in *hȍȍt owl* look like an owl's eyes.

Now see if you can guess the missing letters in *Mad*'s eight zoo-lulus that are shown below and on the next page.

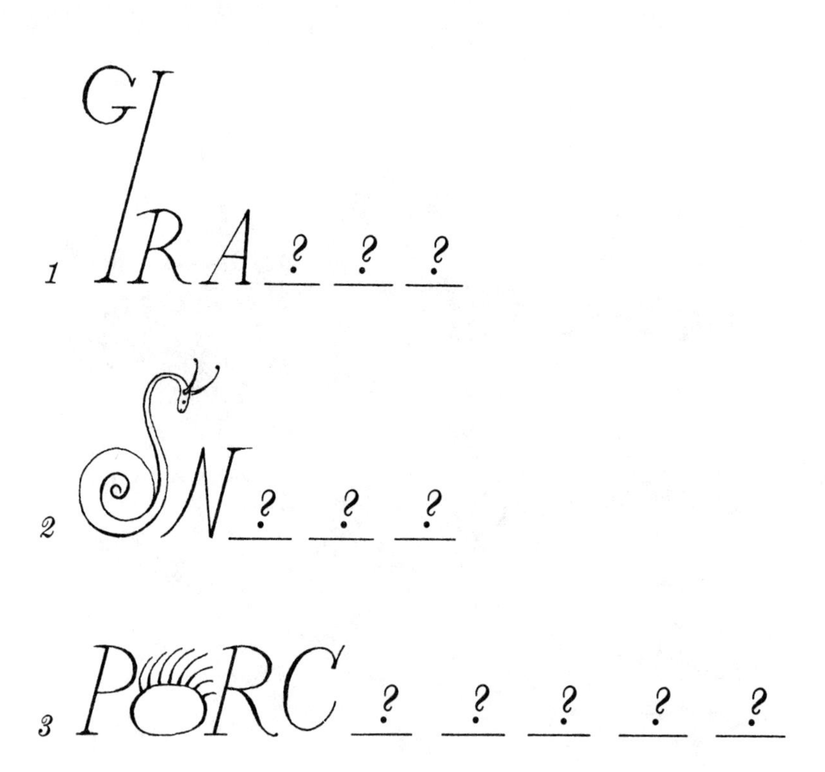

4 SHA ? ?

5 C ?

6 RA ? ?

7 TU ? ? ?

8 B ?

27 *Unscramble the Beast*

The boy at the zoo started to call out the name of the animal he saw, but he became so excited that he got his words all mixed up. See if you can take all twelve of the letters in "Oh, it's a pom pup!" and rearrange them to make a single word that will be the name of the huge beast the boy is pointing to.

Oh, it's a pom pup!

28 *Solve the Animal Equations*

Have you ever added and subtracted letters instead of numbers?

For example, consider the following equation:

1 [bone] − *ONE* + [ear] = ?

The first picture is a picture of a bone, so we print the letters, B-O-N-E. We are told to subtract O-N-E from B-O-N-E, so we cross out O-N-E, which leaves only the B. Next, we must add E-A-R. When we do this, we get the word B-E-A-R. Bear is the animal that solves the equation!

Now see how good you are at this strange kind of arithmetic by working out the following equations. Each one gives the name of a familiar animal.

2 [tie] − *E* + [arm] − [bow] = ?

3 − ***Y*** + − ***M*** = ?

4  +  −  = ?

5 − ***N*** + − + − = ?

29 *The Undecidable Stairway*

Remember that three-pronged, one-slot widget on a previous page? Here's another undecidable figure: a crazy staircase.

If you walk around it clockwise, you can keep on going downstairs forever without ever getting any lower! And if you walk around it counterclockwise, you can keep on forever climbing up the stairs without getting any higher!

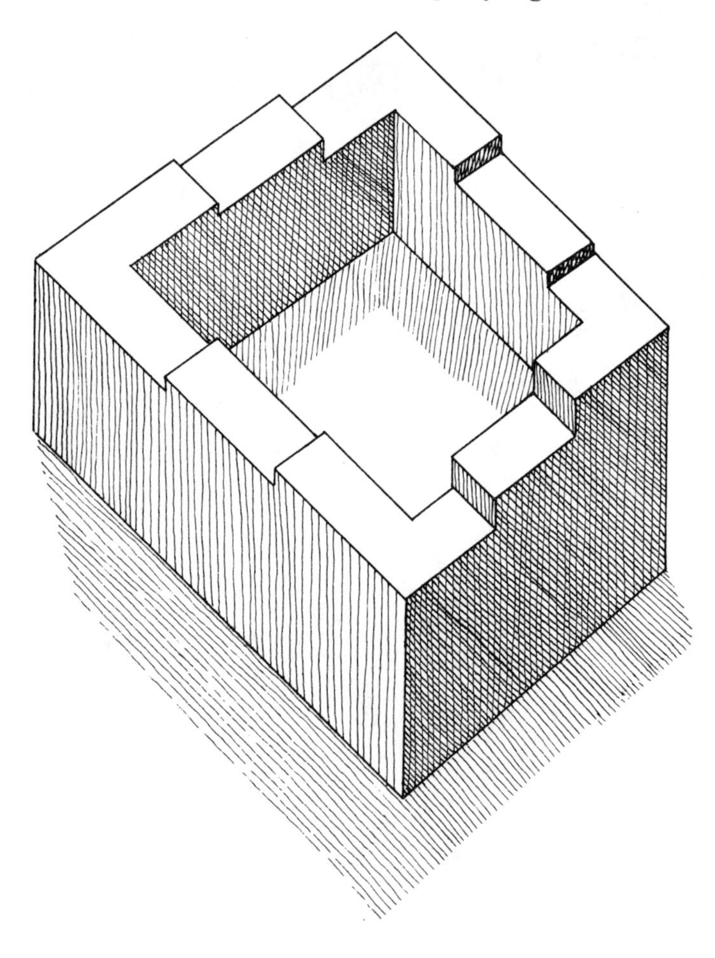

30 *More Sneaky Arithmetic*

1. A harmonica cost a dollar more than a pencil. Together they cost $1.10. How much did each cost?

2. A ribbon is 30 inches long. If you cut it with a pair of scissors into one-inch pieces, how many snips would it take?

3. Farmer Brown came to town with some watermelons. He sold half of them plus half a melon, and found that he had one whole melon left. How many melons did he take to town?

4. If you took 3 apples from a basket that contained 13 apples, how many apples would you have?

5. Nine thousand, nine hundred and nine dollars is written like this: $9,909. How fast can you write the figures for this sum of money: twelve thousand, twelve hundred and twelve dollars?

31 *How Clever are You?*

Are you good at thinking of simple ways to solve difficult problems that you sometimes come up against in everyday life? Suppose, for instance, that:

1. You see a truck that has become stuck beneath an underpass because it was an inch too tall to continue passing through. There is a filling station and garage a short distance down the road. The driver of the truck is starting to walk toward the garage to get help when suddenly a bright idea pops into your head. You tell the driver and five minutes later he is through the underpass and on his way.

What did you tell him to do?

EXPRESS

2. You are a Boy Scout on a hike with your troop. After walking through a small town on your way to Mudville, you reach a spot where two roads cross. A signpost has been knocked over and is lying on its side. None of you knows which road leads to Mudville. Then you remember something that will solve your problem.

What do you remember?

3. You are playing a game of Ping-Pong in the back yard of a friend's house. When you miss the ball, it bounces across the lawn and rolls into a small but deep hole. The hole goes down too far for you to reach the ball with your hand, and the hole bends so much to one side that you can't get the ball by poking a stick into the hole. After a few minutes you think of an easy way to get the ball.

What did you think of?

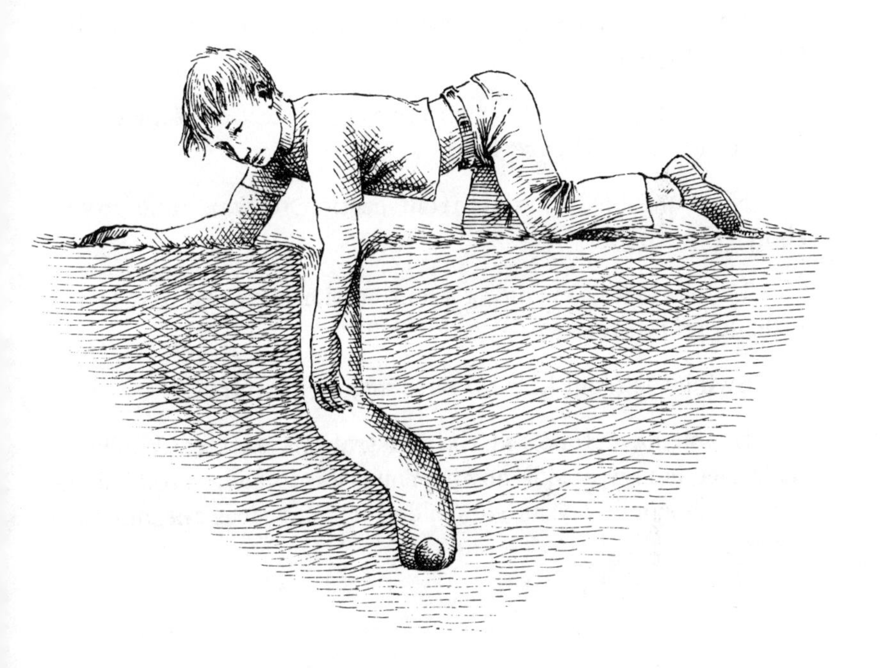

1. TURN GEORGE UPSIDE DOWN

Hold a dollar bill as shown in Figure 1, with Washington's face upright.

Fold down the top half (Figure 2).

Fold in half again, swinging right section back behind left one (Figure 3).

Fold in half again, swinging right section forward in front of left one (Figure 4).

Now unfold, swinging front section forward and to the right (Figure 5).

Unfold again, swinging front section forward and to the right (Figure 6).

Swing the front half forward and up (Figure 7).

If you have followed the illustrations exactly, Washington's face is now upside down! You seem to have unfolded the bill in the same way as you folded it. Why does the bill turn around?

1  WASHINGTON UPRIGHT

2 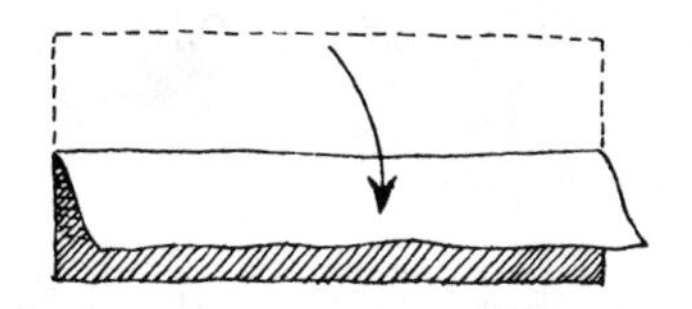 FOLD TOP DOWN

3 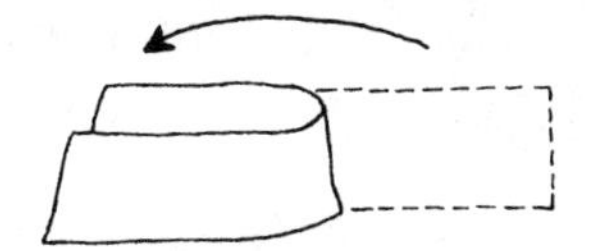 FOLD *BACK* AND TO THE LEFT

4 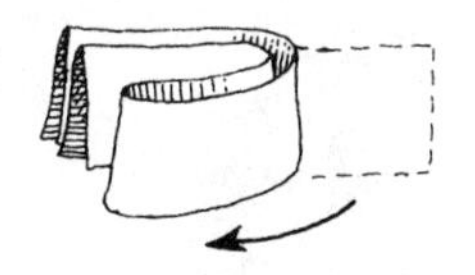 FOLD *FORWARD* AND TO THE LEFT

5 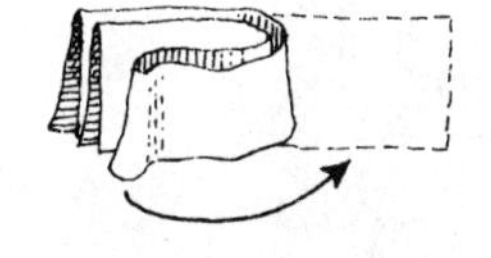 UNFOLD FROM FRONT

6 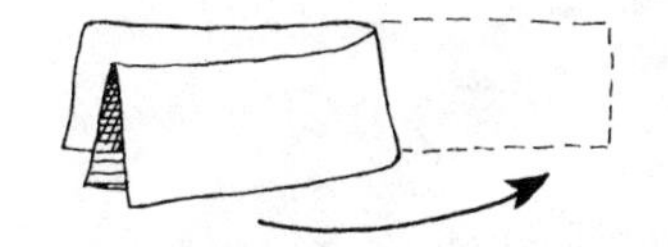 UNFOLD FROM FRONT

7  LIFT UP FRONT FLAP

8  WASHINGTON IS UPSIDE DOWN!

2. TURN GEORGE INTO A MUSHROOM

See if you can make two folds on a dollar bill to turn Washington into a mushroom.

3. FIND GEORGE'S KEY

Inspect the front of a dollar bill carefully. Can you find a picture of a door key?

4. THE POP-OFF CLIPS

Attach two paper clips to a folded dollar bill, exactly as shown in the picture. Hold the ends of the bill and pull the bill out flat. Can you guess what will happen to the clips? Try it and see.

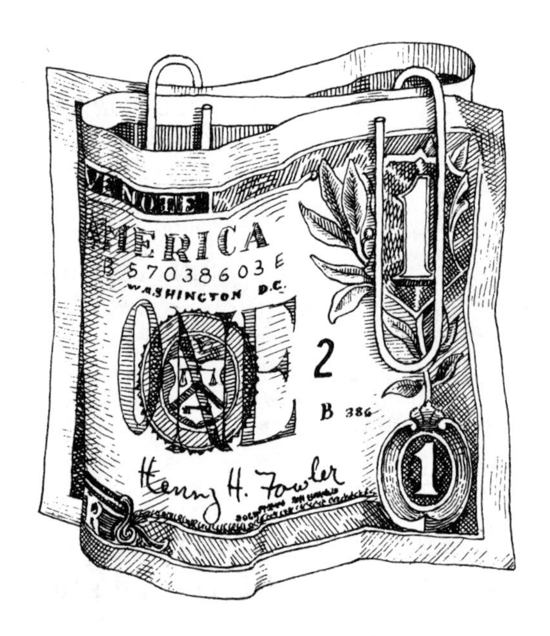

33 *Knock, Knock . . . Who's There?*

You've probably played "Knock, Knock" before, but just in case you haven't, it's a game that goes like this:

"Knock, knock," you say to a friend.

"Who's there?" he replies.

"Gorilla."

"Gorilla who?"

"*Gorilla* my dreams, I love you."

Your friend will laugh (you hope), if he gets the pun. And if he knows a good "knock-knock," *he'll* try it on you.

Before looking at the answers, see how many good "knock-knocks" you can invent for these five boys' names:

1. Hiawatha
2. Sam
3. Noah
4. Tarzan
5. Chester

And these five girls' names:

1. Carmen
2. Sharon
3. Celia
4. Sarah
5. Minnie

34 *Word Bowling*

Mr. Kegler is trying to figure out how he can knock over one bowling pin at a time and always leave pins standing that will spell a word. He has just knocked down the second *T* and the remaining letters spell STARLING, which is the name of a bird. How can he bowl over a pin at a time and each time leave a familiar word, until only one pin that makes a word all by itself is left?

35 *The Great Bracelet Mystery*

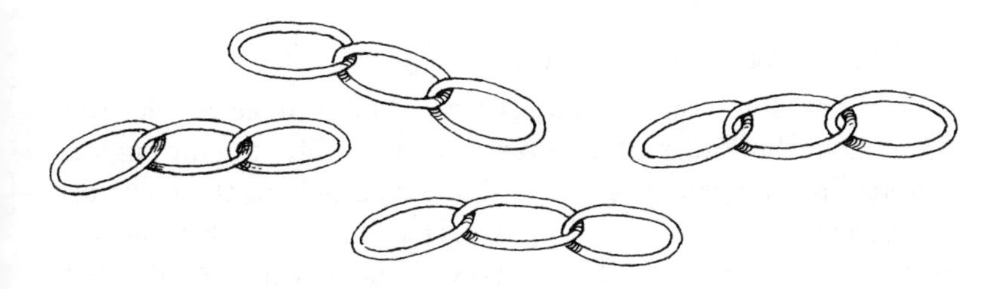

A lady had four pieces of gold chain. Each piece contained three links. She took the four pieces to a jeweler and asked him to join them together to make a bracelet, like this:

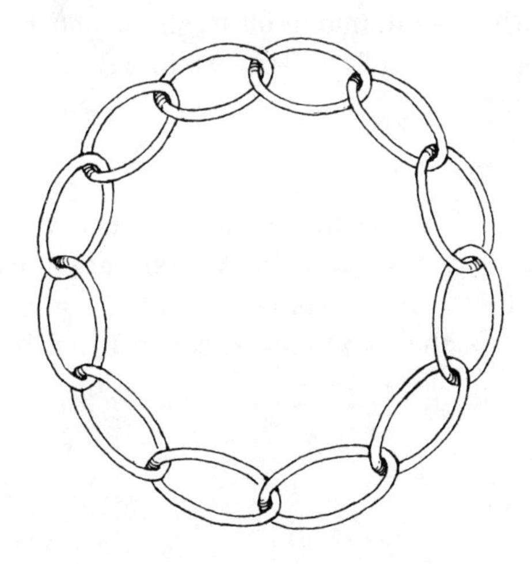

"I'll have to charge you a dollar for each link I cut apart and weld together again," the jeweler said. "Since I have to cut and weld four links, the job will cost you four dollars."

"Oh no it won't," said the lady (who was very good at puzzles). "It should cost only *three* dollars because you can make the bracelet by cutting and welding only *three* links."

The lady was right. Show how the job can be done in just the way she said.

MIDGE ON THE ELEVATOR

Midge lives on the twelfth floor of a modern elevator apartment building. Whenever she gets into the automatic elevator on the ground floor, and no one else is in the elevator, she pushes the button for floor 6, gets off on the sixth floor, and walks up the stairway to the twelfth floor. She would *much* prefer to ride the elevator all the way up to her floor. Why does she do this?

MRS. FUMBLEFINGER'S FUMBLE

Mrs. Fumblefinger was working in the kitchen when a loose ring, with a big diamond on it, slipped off her finger and fell smack into some coffee. Strange to say, the diamond did not get wet. Why?

THE SNEAKY WAITER

At Sloppy Joe's Restaurant a customer was shocked to find a fly in his coffee. He sent the waiter back for a fresh cup. After his first sip, the customer pounded on the table and shouted: "This is the *same* cup of coffee I had before!" How could he tell?

37 *A Pair of Eye Twiddlers*

TALL STILTS

Are the stilts that this clown is standing on bent in or out, or are they straight?

TALL PROVERBS

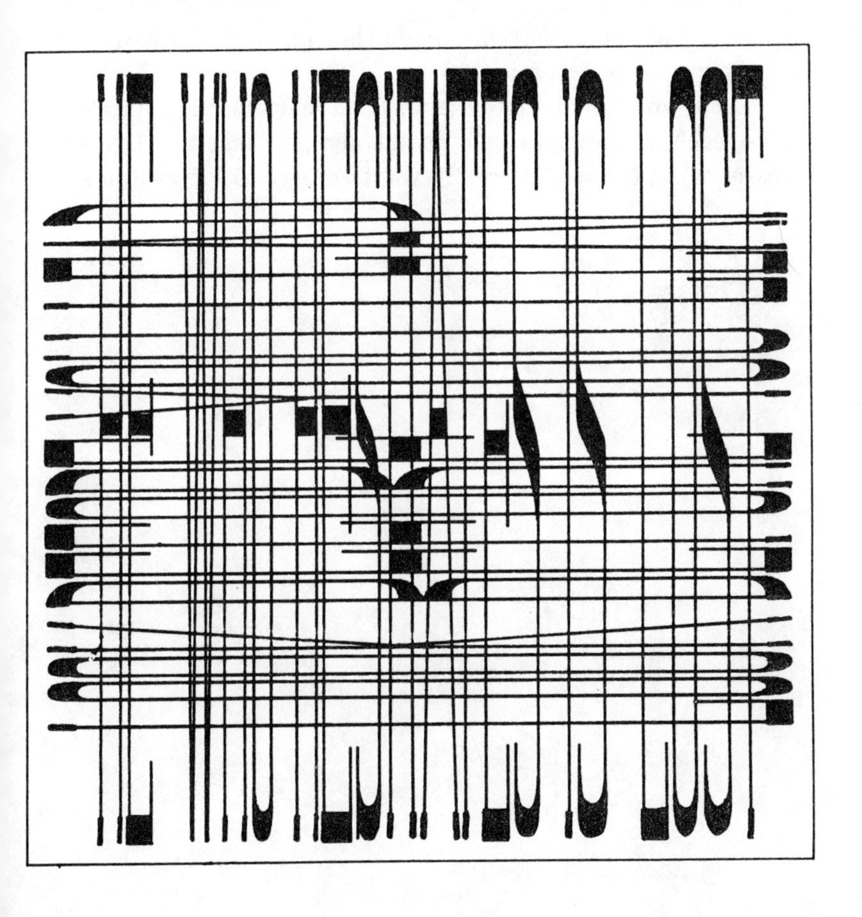

Can you read the two familiar proverbs printed above?

38 *The Puzzling Barbershop Sign*

The owner of this barbershop likes puzzles and jokes, so he put up the sign that you see at the top of his window. When customers come in to ask him for a free shave and drink, he explains that the artist who made the sign for him forgot to add the proper punctuation.

See if you can add one exclamation mark and one question mark, each at the right spot in the sign, so that the sign expresses what the barber really wants to say to his customers.

BARBER
WHAT DO YOU THINK
I'LL SHAVE YOU FOR NOTHING
AND GIVE YOU A DRINK

39 *The Amazing Mr. Mazo*

Start at Mr. Mazo's necktie and find a way to reach the inside of his hat without crossing any lines. Use your finger or a toothpick, instead of a pencil, to follow the route, so you can show the puzzle to your friends without giving away the answer.

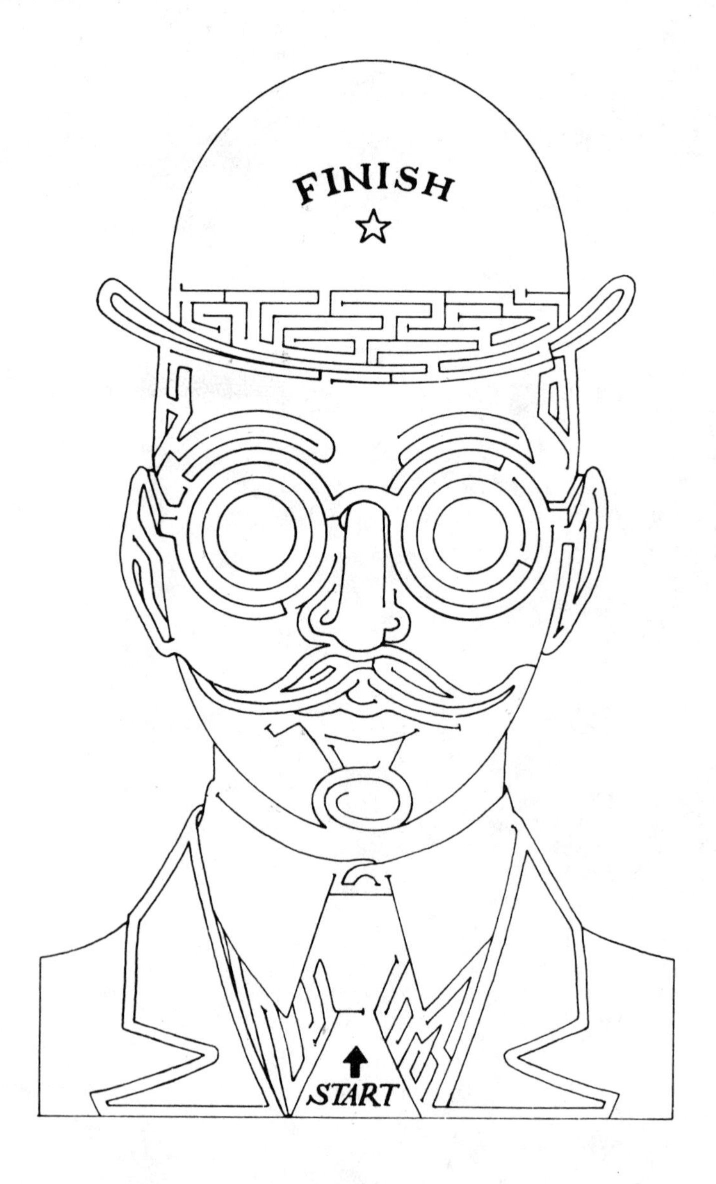

Old Mother Hubbard
Went to the cupboard
To get her poor dog a bone.
When she got there,
The cupboard was bare,

and so Old Mother Hubbard cried out, "OICURMT!"

Can you figure out the meaning of that strange exclamation?

41 *Solve the Bird Equations*

The equations on these two pages work the same way as the animal equations on previous pages except that each gives the name of a familiar bird. See how many of them you can solve.

1

— + = ?

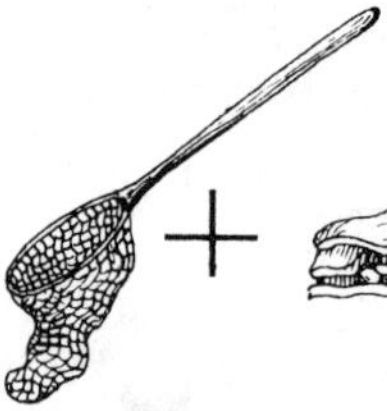

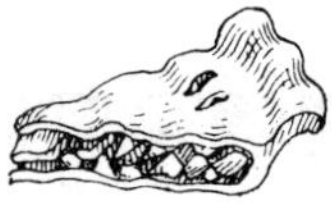

2

— + + — BEEF = ?

3 − + = ?

4 + − +

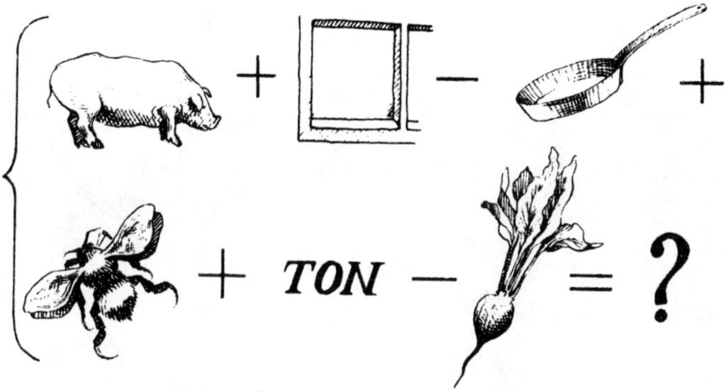

+ TON − = ?

5 + + − = ?

THE KING AND THE ALCHEMIST

One day, centuries ago, an alchemist brought a small bottle to the king. "This bottle," said he, "holds a liquid so powerful that it will instantly dissolve anything it touches." How did the king know the man was lying?

BETSY AND PATSY

"We were born on the same day of the same year," said Betsy.

"And we have the same mother and father," said Patsy.

"But we're not twins," said Betsy.

Can you explain?

THE PURPLE PARROT

"I guarantee," said the salesman in the pet shop, "that this purple parrot will repeat every word it hears." A customer bought the bird, but found that the parrot wouldn't speak a single word. Nevertheless, what the salesman said was true. How could this be?

43 *The Concealed Proverb*

In each of the sentences below a word is concealed, such as the word "no" that is marked in the fifth sentence. If you can find the other buried words and read them in order, they will form a well-known proverb.

1. The word buried here has only one letter.
2. Did you find a jelly roll in Gaskin's Bakery?
3. It's the best one I've ever seen.
4. The rug at her stairway was made in India.
5. He's a<u>n o</u>ld friend.
6. Amos sold his bicycle to a friend.

44 *The Dime-and-Nickel Switcheroo*

Put three pennies, a nickel, and a dime on top of their pictures. By sliding one coin at a time, into a neighboring empty cell, can you make the dime and nickel change places? You are allowed to move a coin left or right, up or down, but not diagonally.

It's easy to switch the dime and nickel if you keep sliding the coins long enough, so try to figure out how to switch them with the smallest number of moves. It can be done in fewer than twenty moves, but it takes more than twelve.

45 *Mrs. Windbag's Gift*

Professor Windbag likes to use big words and say everything in the most complicated way he can. When he handed a birthday present to his wife he said:

"My dear, here is a diminutive, aurum, truncated cone, convex on its summit and semiperforated with symmetrical indentations and a hollow interior."

Can you guess what is in the box?

46 *More Eye Twiddlers*

CRAZY LINES

Which line is longest: the line from *A* to *B* or the line from *A* to *C*?

CRAZY CURVES

Are these curves circles or ovals?

47 *Shake Out the Cherry*

The matches in this picture represent a cocktail glass with a cherry inside. Put a paper match on top of each match in the picture.

Now see if you can pick up just two matches and place them so that the cocktail glass is upside down, and the cherry is *outside* the glass.

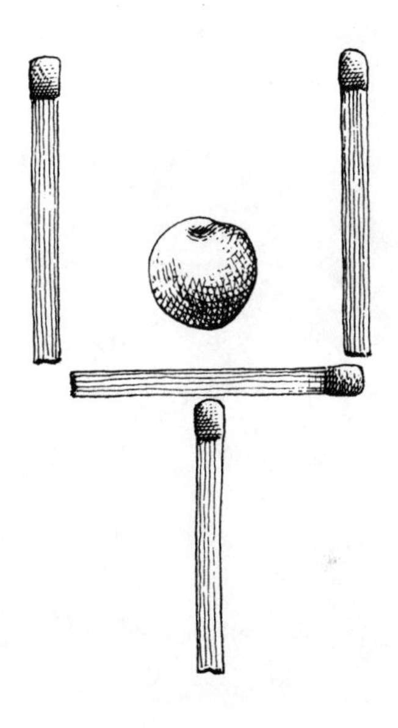

ANSWERS

1 *Ridiculous Riddles*

1. Super pickle. 2. A magnetized banana. 3. Here, kitty, kitty, kitty. 4. An embarrassed zebra. 5. Dwight D. Eiffel Tower. 6. With some root beer, two scoops of ice cream, and a hippopotamus. 7. *Very* carefully. 8. A lost camel. 9. A cherry that works at night as a grape. 10. It's the part of an apple you throw away. 11. A chair, a bed and a toothbrush. 12. A remote-control fig. 13. A baseball team with the measles. 14. Because they can't walk at all *hardly*. 15. Nothing. 16. A cow. 17. Tep on the brake, toopid! 18. He just ate a ranger.

2 *Handies*

1. Indian peeking over indoor television antenna. 2. Absent-minded professor scratching his head. 3. Midget playing the piano. 4. Help! I swallowed my toothbrush! 5. Tea (T) for two. 6. Turn hand upside down, with fingers in same position, and say: It's a dead *that*. 7. Helicopter looking for a place to land.

3 *Fun with Palindromes*

There are twelve palindromic words. Taken in order they are: radar, Hannah, I, redder, did, noon, wow, ma'am, Otto, a, pup, eye.

4 *The Lost Star*

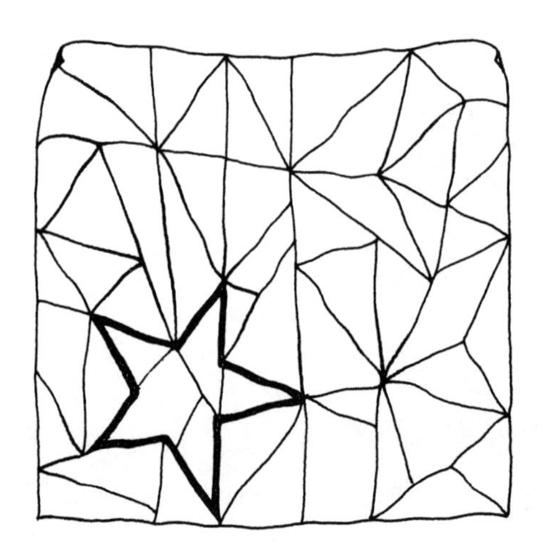

5 *Find the Hidden Animals*

1. Dog. 2. Gnu. 3. Monkey. 4. Beaver. 5. Bear. 6. Lion. 7. Camel 8. Cat.

6 *Tricky Questions*

HIGGS'S PIGS

None. Pigs can't talk.

PENNIES FOR SALE

One thousand, nine hundred and sixty-six pennies are worth $19.66, which is almost twenty dollars.

POP AND GRANDPOP

Yes. Strange as it may seem, Tom's father is forty and his grandfather, on his *mother's* side, is forty-six. The grandfather was twenty when Tom's mother was born and she was sixteen when Tom was born. Tom is now ten years old. ($20 + 16 + 10 = 46$.)

It's easy. First Jim crawls through the pipe in one direction. After he comes out, Tom crawls through it the other way.

7 *The Five Airy Creatures*

The little "creatures" are the five vowels: *A,E,I,O,U*. Jonathan Swift called them "airy" because they are actually made of air—the air that comes out of your throat and makes the vowel sounds.

8 *The Maze of the Minotaur*

The path to the Minotaur is shown by the dotted line.

START HERE

9 *The Dime-and-Penny-Switcheroo*

1. Slide the penny.
2. Jump the penny with the dime.
3. Slide the dime.
4. Jump the dime with the penny.
5. Jump the other dime with the other penny.
6. Slide the dime.
7. Jump the penny with the dime.
8. Slide the penny just jumped.

10 *A Dozen Droodles for Nimble Noodles*

1. Toothbrush with only two bristles.
2. A boy who needs a haircut.
3. The Wicked Witch of the West after Dorothy tossed a pail of water over her.
4. Four elephants sniffing a baseball.
5. Navel orange wearing a bikini.
6. Baby sleeping in dresser drawer.
7. Three little pigs on a foggy day.
8. A used lollipop.
9. Elephant scratching an ant's back.
10. Hard-fried egg turned upside down on a counter.
11. Man with bow tie that got caught in elevator doors.
12. Banana skipping rope.

11 *Tantalizing Toothpick Teasers*

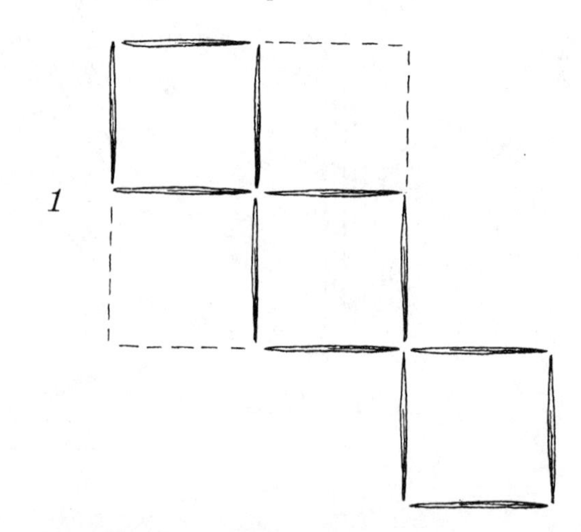

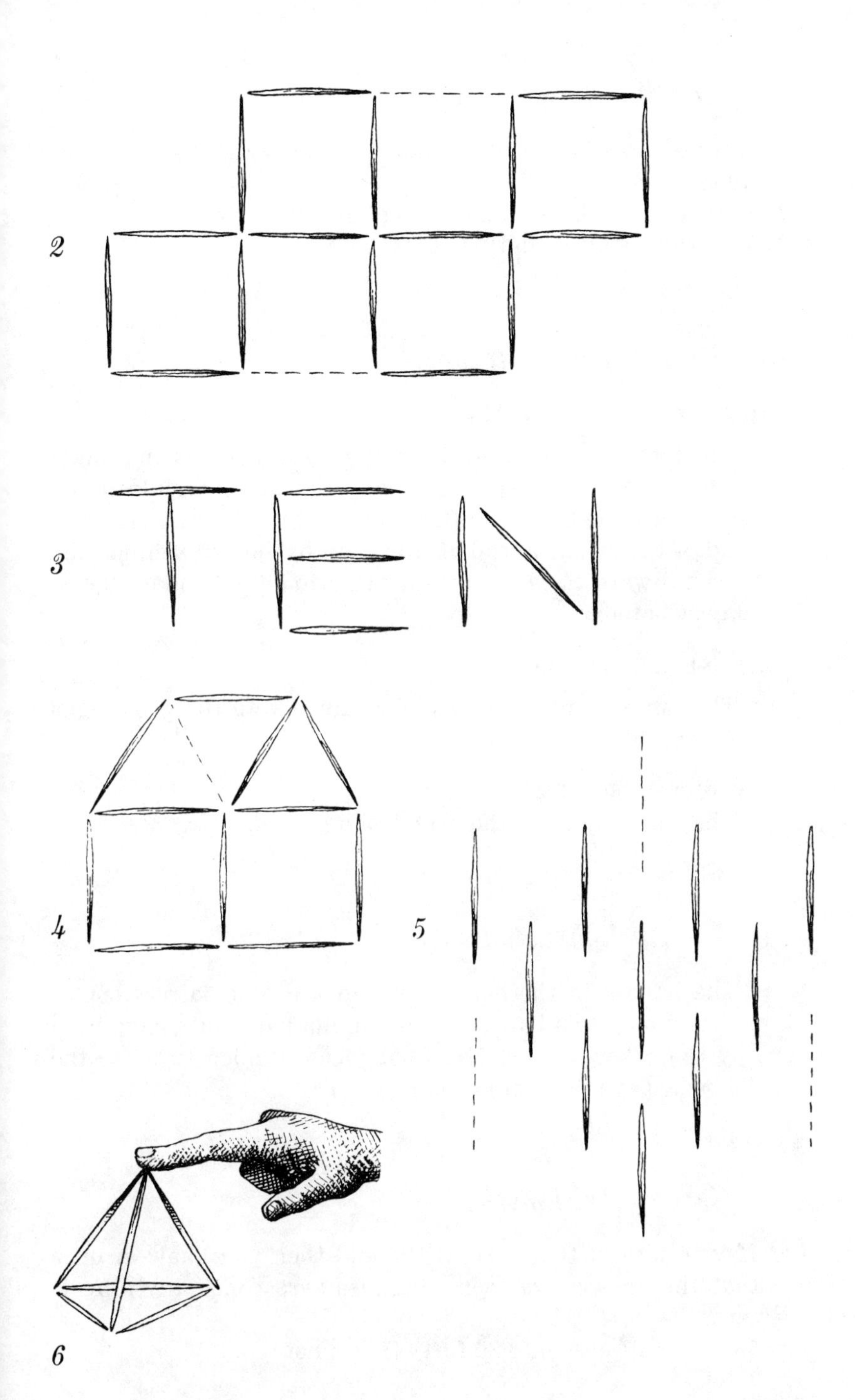
2
3
4
5
6

12 *Read the Thriftigrams*

1. I'm never as happy as I am when I'm with you. Love and kisses.
2. Have to take plane. Can you meet me at airport?
3. Will you be my Valentine? I love you.

13 *More Tricky Questions*

THE TRAMP AND THE TRAIN

The tramp had been walking along train tracks on a high, narrow bridge with no space on the sides where he could stand. When he saw the train speeding toward him, he was closer to the end of the bridge ahead of him than to the end behind him, so he ran toward the approaching train to get off the bridge as quickly as possible.

A HARD-BOILED PROBLEM

The same twenty minutes. You can put all the goose eggs in the same pan.

HEAP TOUGH PROBLEM

The big Indian was the little Indian's *mother.*

14 *Mr. Bushyhead's Problem*

The barber in the neat shop had the best haircut. Since there were only two barbers in town, his hair must have been cut by the *other* barber. Mr. Bushyhead decided to go to the barber who gave the best haircut.

15 *Sneaky Arithmetic*

1. If you noticed that zero at the end then you knew at once that the answer was zero, because zero times anything is zero.
2. 43. One half goes into 20 forty times, not ten.

3. 10 pounds. Two halves make a whole, so if the brick's total weight is the sum of 5 pounds and half the total weight, the other half must also be 5.
4. If all but 9 died, then of course 9 were left.
5. Twice one half of any number is that same number, so the answer is 987,654,321.

16 *The Careless Sign Painter*

The signs should read: QUIET, PUSH, EXIT, and WET PAINT.

18 *Sally's Silly Walk*

Last Sunday, when Sally went for a walk, she saw a policeman; skipping rope, she saw a fire engine; eating an ice cream cone, she saw a squirrel; humming a tune, she saw a puppy; climbing a tree, she saw two robins; playing hopscotch, she saw an organ grinder and his monkey.

19 *Still More Tricky Questions*

THE MISPELLED WORD

The word that is spelled incorrectly is the word "mispelled" in the problem's title. There should be another "s" in misspelled.

FLAPDOODLE'S WALK

Archibald was bald.

STAMPS TO STUMP YOU

Twelve. It takes twelve of anything to make a dozen. Even four-cent stamps.

WHAT DO YOU THINK?

The poem doesn't ask a question. The horse's name was *What-do-you-think.*

20 *Guess The Typitoons*

1. Boy watching baseball game through a hole in the fence. 2. Person watching a Ping-Pong game. 3. Row of soldiers (or cello players). 4. Row of bunnies (or bugs). 5. The leaning tower of Pisa. 6. Boy in third row raising his hand to ask teacher a question.

21 *The Fish and the Robot*

There are many ways to draw both figures according to the rules, but in every case you must always begin and end the single line at the spots marked with dots.

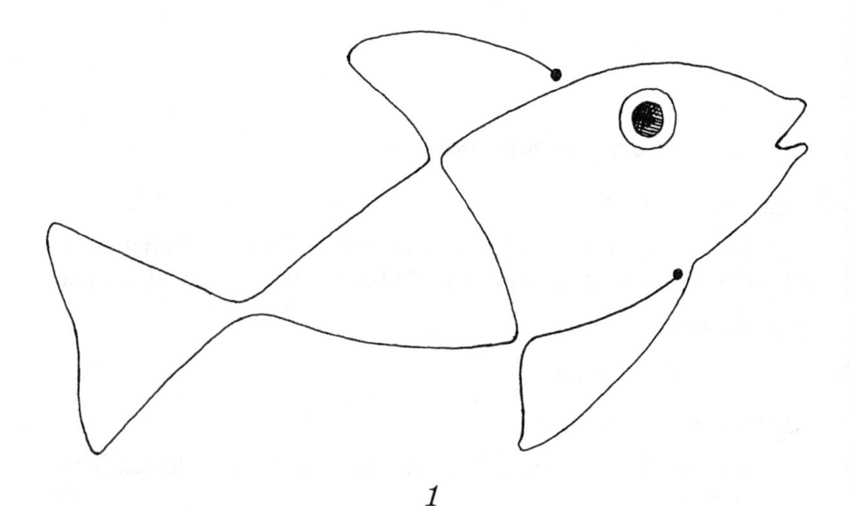

1

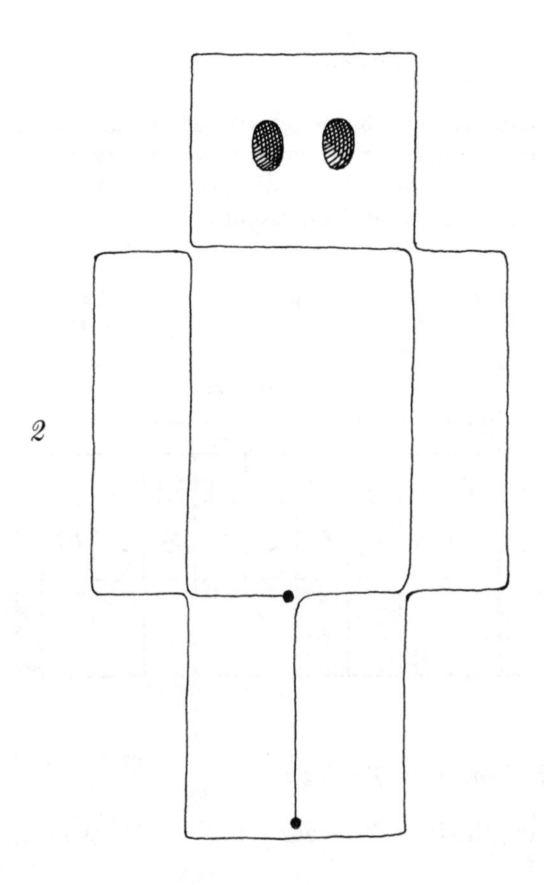

22 *The Christmas-Star Puzzle*

The secret: Put the first penny on any point you like and slide it to another point. After that, place each penny so you can slide it to the spot where the previous penny was *before* you slid it. For example:

1. Put a penny on A and slide it to C.
2. The previous penny was on A before you slid it, so put a penny on D and slide it to A.
3. The previous penny was on D before you slid it, so put a penny on B and slide it to D.
4. The previous penny was on B before you slid it, so put a penny on E and slide it to B.

23 *The Boring Bookworm*

The worm travels 16 inches. When a book stands on a shelf in front of you, its front cover is on the *right* side and its back cover is on the *left*. The worm travels, therefore, along the 16-inch path shown by the dotted line.

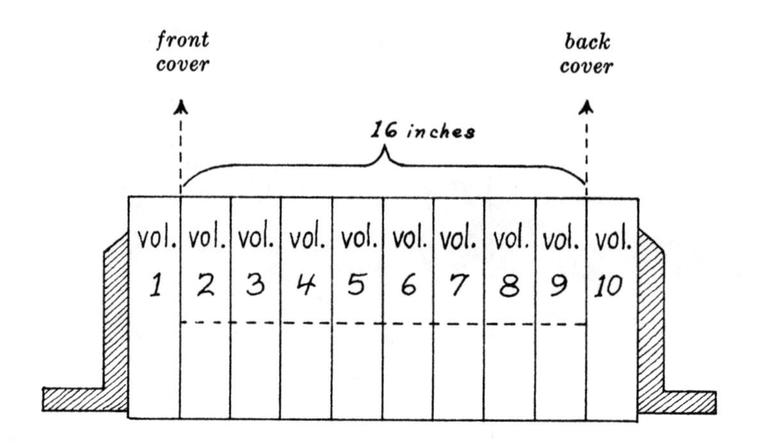

24 *The Triangular Turkey*

There are thirteen different triangles. No wonder it's an unlucky bird!

25 *Find the Best Words*

Abominable (a bomb in a bull) and *Noble* (no bull).

26 *Zoo-lulus*

Giraffe, snail, porcupine, shark, cow, rabbit, turkey, bat.

27 *Unscramble the Beast*

Hippopotamus.

28 *Solve the Animal Equations*

2. TIE − E + FINGER − FIN = TIGER
3. BABY − Y + MOON − M = BABOON
4. BADGE + PEAR − PEA = BADGER
5. CAN − N + MICE − ICE + ELBOW −BOW = CAMEL

30 *More Sneaky Arithmetic*

1. The harmonica cost $1.05, the pencil cost 5¢. Perhaps you thought the harmonica cost $1 and the pencil cost 10¢, but then the harmonica would cost 90¢ more than the pencil. To cost a dollar more, the harmonica must cost $1.05 and the pencil 5¢, because $1.05 minus 5¢ is $1.
2. Twenty-nine snips. The last two inches are divided by one snip.
3. Farmer Brown originally had 3 watermelons. He sold half of them (1½) plus half a melon, which is the same as saying that he sold 2 melons, leaving him with one whole melon as stated.
4. If you took 3 apples, you would *have* 3 apples!
5. $13,212. (12,000 + 1,200 + 12 = 13,212.)

31 *How Clever are You?*

1. You tell the driver to let some air out of his tires. This lowers the truck enough to let it through the underpass, then the driver can stop at the garage ahead and put the air back in his tires.
2. You remember the name of the town you have just left. If the signpost is replaced in the hole, with the name of the town you have left pointing back along the road you have just traveled, all the other signs will have to be pointing in the right direction.
3. Fill the hole with water from a hose and the Ping-Pong ball will float to the top.

32 *Folding-Money Fun*

1. TURN GEORGE UPSIDE DOWN

The secret is in the fold shown in the third picture. Note that it goes *back* instead of forward. When you undo this fold, however, as shown in the sixth picture, you undo it from the *front*. This is what turns the bill around. Practice until you can do the folds fast, without thinking, and you can have fun showing this to your friends. When *they* try it, the bill stays right side up!

2. TURN GEORGE INTO A MUSHROOM

3. FIND GEORGE'S KEY

The key is inside the round green seal at the right of Washington's picture.

4. THE POP-OFF CLIPS

The clips pop into the air and fall linked together!

33 *Knock, Knock. . . . Who's There?*

Boys' names: 1. *Hiawatha* high school dropout. 2. *Sam* enchanted evening. 3. *Noah* body knows the trouble I've seen. 4. *Tarzan* stripes forever. 5. *Chester* minute and I'll try to find out.

Girls' names: 1. *Carmen* get it! 2. *Sharon* share alike. 3. *Celia* later. 4. *Sarah* doctor in the house? 5. *Minnie* brave hearts are asleep in the deep.

Did you think of any better ones?

34 *Word Bowling*

STARTLING
STARLING
STARING
STRING
STING
SING
SIN
IN
I

35 *The Great Bracelet Mystery*

To make the bracelet without cutting more than three links, simply open all three links of one piece, then use those three links to join the other three links into a circle, like this:

36 *And Still More Tricky Questions*

MIDGE ON THE ELEVATOR

Midge is a small child and can reach up only as high as the sixth-floor button.

MRS. FUMBLEFINGER'S FUMBLE

The ring fell into a can of dry, ground coffee.

THE SNEAKY WAITER

The man had put sugar in his coffee before he found the fly in it. When he tasted the sugared coffee, he knew that the waiter had brought back the same cup of coffee.

37 *A Pair of Eye Twiddlers*

TALL PROVERBS

Hold the page flat in front of your nose and sight along it with one eye. (Be sure to keep the other eye closed.) You should be able to read the first proverb: *He who hesitates is lost.*

Now turn the book so you can sight along the page, the same way as before, but from the right side. You should be able to read the second proverb: *Look before you leap.*

The proverbs give opposite advice. Which do you think is the best to follow?

TALL STILTS

The stilts are perfectly straight! You can prove it by putting the edge of a ruler along them or by holding the page flat in front of one eye and looking at the stilts in the same way that you looked at the proverbs.

38 *The Puzzling Barbershop Sign*

WHAT! DO YOU THINK I'LL SHAVE YOU FOR NOTHING AND GIVE YOU A DRINK?

39 *The Amazing Mr. Mazo*

The dotted line shows how to go from Mr. Mazo's necktie to the inside of his hat.

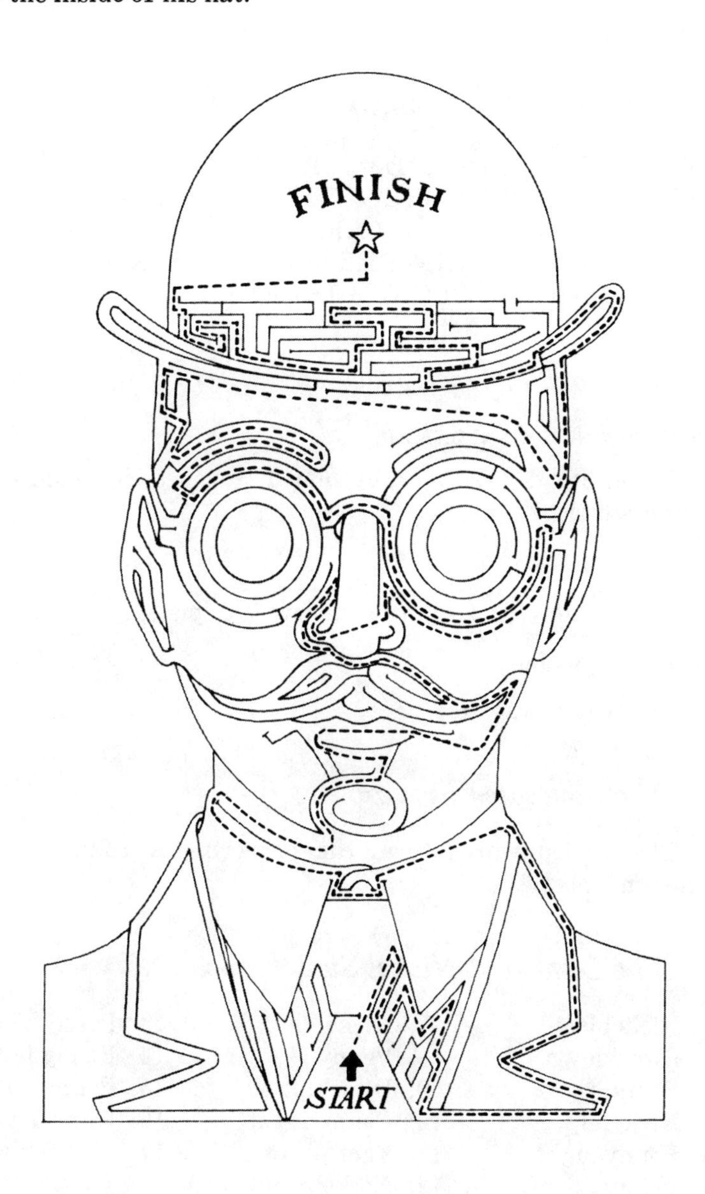

40 *Mother Hubbard's Cupboard*

Read the letters of Mother Hubbard's exclamation out loud and you will hear yourself saying "Oh, I see you are empty!"

41 *Solve the Bird Equations*

1. MAGNET – NET + PIE = MAGPIE
2. STRING – RING + BEE + FORK – BEEF = STORK
3. LACE – ACE + ARK = LARK
4. PIG + PANE – PAN + BEE + TON – BEET = PIGEON
5. STAR + TREE + SLING – TREES = STARLING

42 *The Last Tricky Questions*

THE KING AND THE ALCHEMIST

If the liquid dissolved anything it touched, it would dissolve the bottle.

BETSY AND PATSY

Betsy, Patsy, and a third sister were triplets.

THE PURPLE PARROT

The parrot was deaf.

43 *The Concealed Proverb*

The buried words form the proverb, "A rolling stone gathers no moss."

44 *The Dime-and-Nickel Switcheroo*

1. Nickel left
2. Dime down
3. Penny right
4. Nickel up
5. Penny right
6. Penny down
7. Nickel left
8. Penny left
9. Dime up
10. Penny right
11. Penny down
12. Nickel right
13. Penny up
14. Penny left
15. Penny left
16. Dime down
17. Nickel right

45 *Mrs. Windbag's Gift*

Mr. Windbag is saying that he is giving his wife a gold thimble.

46 *More Eye Twiddlers*

CRAZY LINES

The two lines are the same length, as you can prove by measuring them.

CRAZY CURVES

The curves are circles. You can prove this by measuring the distances from the central spot to different points on the same curve. You'll find that on any curve, all points along it are the same distance from the center.

47 *Shake Out the Cherry*

The two matches are moved like this:

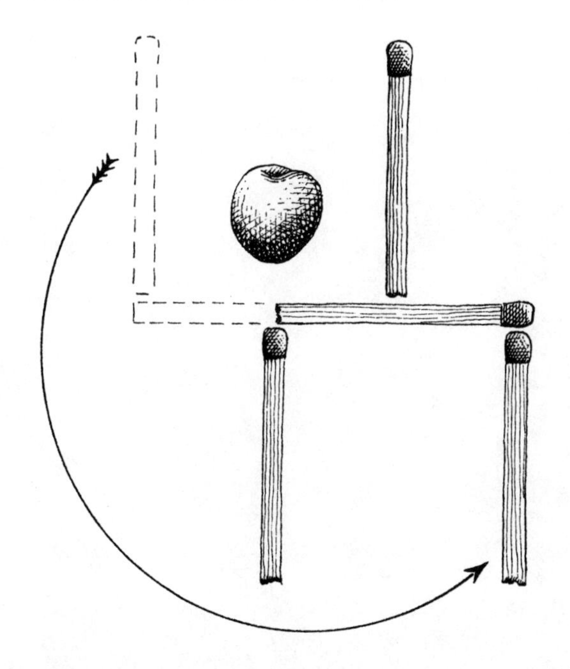

MORE PERPLEXING PUZZLES *and* TANTALIZING TEASERS

For my grand-nephew
Bobby Schoeppel

Contents

Puzzle Number

Introduction

Dear Friend:

This book is a sequel to *Perplexing Puzzles and Tantalizing Teasers*. The same artist who illustrated that book, Mr. Laszlo Kubinyi, has illustrated this one. Don't you agree that he has caught the spirit of the puzzles and done a marvelous job?

If you liked the previous book, you are sure to like this one. Its puzzles are on the same level of difficulty, and they have been carefully selected to exercise your brain in a way that I guarantee you won't find boring. The solutions are at the back of the book. I can't stop you from peeking at those answers, but I can tell you that, if you resist the temptation until you've tried as hard as you can to solve a puzzle, you'll get much more pleasure from the book.

Please write to me, in care of the publishers, if you have any good ideas for new puzzles, or to tell me which puzzles you liked best.

Martin Gardner

1

Ten Ridiculous Riddles

1. What does a duck do when it flies upside down?
2. If all cars in the United States were pink, what kind of flower would the United States be?
3. How do you keep a skunk from smelling?
4. What did George Washington say to his men before they got in their boat to cross the Delaware?
5. Why does Tony the fireman have red suspenders?
6. What's big, green, and loves hot tamales?
7. Where do vampires get their mail?
8. What's Smokey the Bear's middle name?
9. What do you get when you cross a hippo with a jar of peanut butter?
10. What's gray, has four legs, a tail, and a trunk?

2

Fat Bats and Other Funny Beasts

Each picture shows an animal that can be described by two words that rhyme. Two examples are given. Can you find a rhyming pair of words for the other animals?

Fat bat

Cryin' lion

3

"The Whistler"

Mr. and Mrs. Smith have a son who is an artist. He has just sent them a painting. In a letter he said that the picture's title is "The Whistler."

Mr. and Mrs. Smith are puzzled. All they can see is a circle and four spots. Why in the world did their son call it "The Whistler"? And what does that label "68 ON" mean?

4

Where Does the Ball Go?

Keep your eyes on that tiny "x" between the tennis ball and the boy's face. Then bring the page slowly toward the tip of your nose. You'll see the ball fly straight into the boy's mouth!

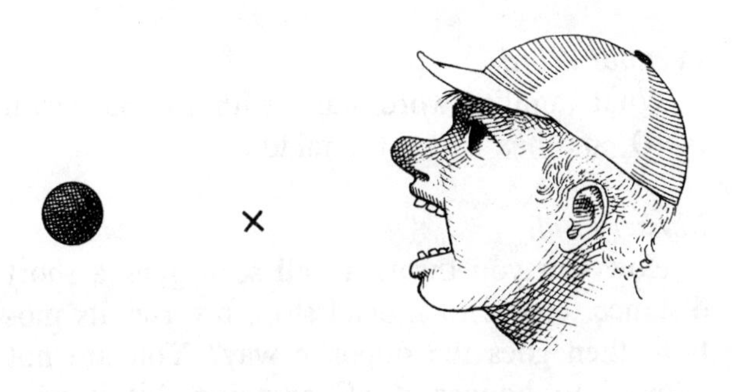

5

Tricky Questions

Speedy Retirement

Bascom turned off the light in his bedroom and was able to get to bed before the room was dark. His bed is 15 feet from the wall switch. How did Bascom do it?

Peculiar Word

What familiar word starts with IS, ends with AND, and has LA in the middle?

Mystery Ball

How can you throw a ball so it goes a short distance, comes to a dead stop, reverses its motion, then goes the opposite way? You are not allowed to bounce it off anything, hit it with anything, or tie anything to it.

Shrewd Barbers

Why do barbers in Los Angeles prefer cutting the hair of ten fat men to cutting the hair of one skinny man?

6

What Do You Do Next?

Here are a dozen funny jokes to play on friends. In every case the joke is doing something your friends don't expect. Pretend you are playing these jokes on someone. See if you can guess what you are supposed to do after your friends respond to your first remark.

1. Did you know I can stick out my tongue and touch my nose?
2. Brush the back of your hand lightly down over a person's face, from forehead to chin, and say, "Did you like that?"
3. Do you know how to make an ordinary pencil, with black lead, write red or blue?
4. Put the palm of your right hand on top of your left fist, and then say "wing" three times.
5. Hand me your plate and I'll show you how to push it through the handle of my cup.
6. I can drop this paper match on the table so it will land on its *edge* and stay there.

7. I can have you and a friend stand on the same sheet of newspaper in such a way that neither of you can touch the other.

8. Would you like to see me crawl into a pill bottle?

9. Would you like to have your palm read?

10. Ever see a match burn underwater?

11. What's sexy and hums?

12. Do you know how to keep an idiot in suspense?

7

The Musicians of Inviz

Inviz is one of the strangest of all villages in Oz. Every day in Inviz all objects of a certain type become invisible. The pictures show what happened on the day that no one could see a musical instrument. Your task is to guess what instrument is being played by each musician. (See art on pp. 10-11).

8

To Be or Not to Be

There are at least 89 things in this picture with names that begin with the letter B. How many can you find?

MEA
AIRCUT

9

What's the Difference?

The picture on the right page shows the same scene as the picture on the left, except it has been reflected in a mirror. Aside from that, however,

the two pictures are not exactly alike. If you look carefully, you'll find six spots where things are not the same. Can you find all six?

10

Easy as ABC

1. The design below is for a Christmas card. What single-word Christmas greeting does it convey?

A	B	C	D	E
F	G	H	I	J
K	M	N	O	P
Q	R	S	T	U
V	W	X	Y	Z

2. The artist has divided all capital letters into two groups: those on the line and those dropped below the line. On what basis did he make the division?

A EF HI KLMN T VWXYZ
BCD G J OPQRS U

3. The word "moon" can be formed by using letters next to one another in the alphabet as shown by the bracket. How many dictionary words can you form by using adjacent letters?

We won't count the obvious single-letter words "a" and "I," or easy two-letter words such as "hi," "on," and "no." There are more than 20 words of three or more letters that you can make, most of them familiar words. Remember, a letter can be used more than once, but all the letters must be next to one another in the alphabet.

MOON

ABCDEFGHIJKL M N O PQRSTUVWXYZ

11

Four Eye Twiddlers

The Two Spirals

One of these spirals is formed with a single piece of rope that has its ends joined. The other is formed with two separate pieces of rope, each with joined ends. Can you tell which is which by using only your eyes? No fair tracing the lines with a pencil.

The Ghost Triangle

Doesn't it look as if a white triangle is pasted on the page? The white triangle is an illusion. It isn't there at all.

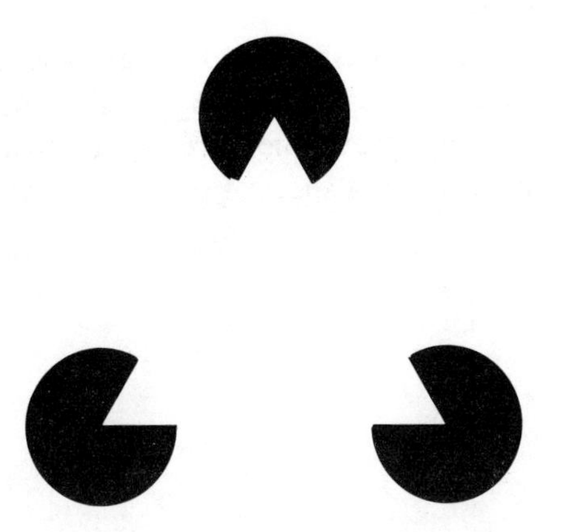

Trap the Beetle

The beetle seems to be on the outside of the box. But stare at the box for a minute or two and something strange will happen. It sort of turns inside out and you'll see the beetle on the *inside,* on a checked floor of the box.

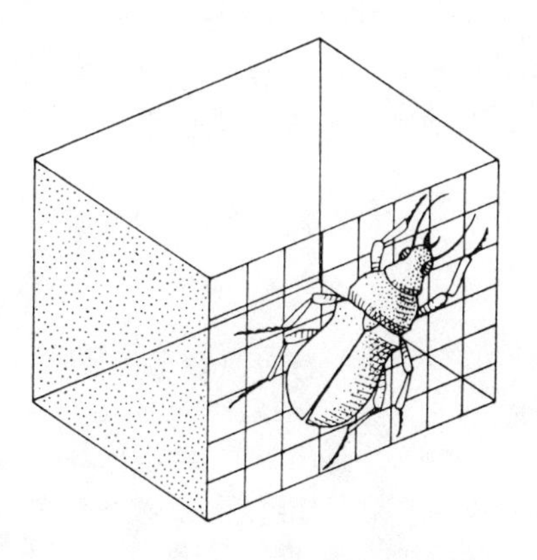

The Curious Cube

There are two spots in this picture where a pair of line segments meet at right angles. Can you find them? You can prove they are right angles by fitting the corner of a sheet of paper into them.

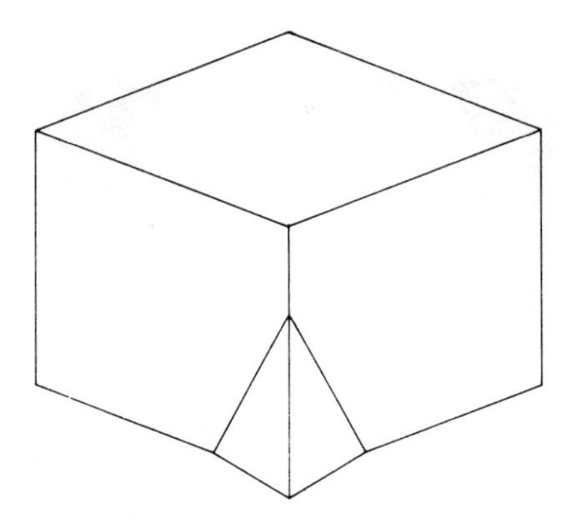

12

Dividing the Cake

One large piece of cake is to be shared between Henry and his sister Henrietta. The two are squabbling over who gets to cut the cake. Each thinks the other will cut it so as to give himself or herself the larger portion.

Dad let them argue for a while before he said, “May I make a suggestion? I'll show you how to decide the matter so both of you will be completely satisfied with your share.”

When Henry and Henrietta heard Dad's clever scheme, they agreed it would work. What did Dad suggest?

13

Find the Mistakes

The artist has made 27 mistakes in this picture, but some are not easy to find. For instance, the keyhole in the door is upside down. See how quickly you can spot the others.

TREASURE ISLAND

14

Three Match Puzzles

1. Arrange four paper matches as shown. Move just one match to make a square.

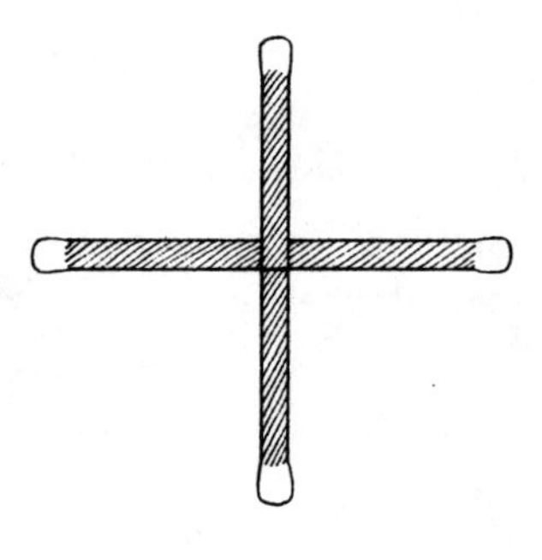

2. Nine matches make the equation below. The numbers are in Roman numerals. The equation is wrong because 1 minus 3 does *not* equal 2. Move just one match to correct the equation.

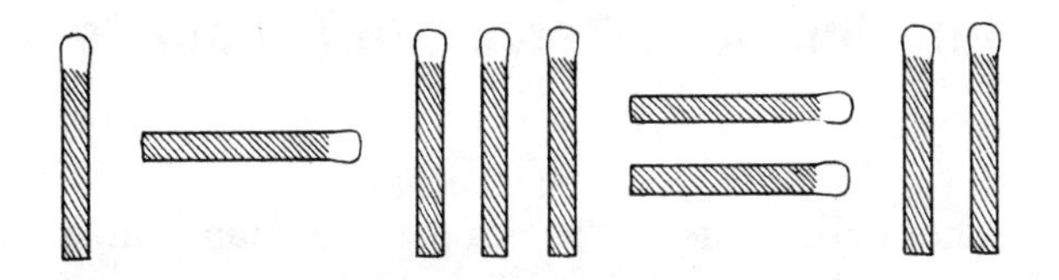

3. Here's another equation of Roman numerals, made with ten matches. It, too, is incorrect. Can you correct the equation without touching the matches, adding new matches, or taking away any matches?

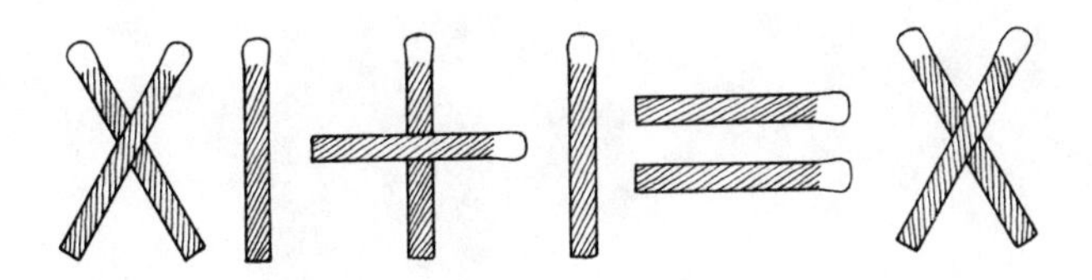

15

Help Sherlock Catch the Criminals

Sherlock Holmes is looking for a man and a woman who are wanted by the police. He's been told they are somewhere in this airline terminal, pretending not to know each other. The man is clean shaven, with dark hair, no glasses, a bowtie and a light-colored suit. The lady is a blonde, wearing a scarf and glasses, and carrying a light-colored shoulder bag. How quickly can you find them?

16

The Two Watering Cans

A woman is trying to decide which of the two watering cans to buy. She wants the one that holds the most water. Which one do you recommend?

17

See Sherman Shave

Sheldon Sherman has been on a fishing trip and hasn't shaved for five days. Here you see him halfway through shaving.

To see how Sheldon looked *before* he started shaving, place the edge of a mirror on the dotted line and look into the mirror from the left.

To see how Sheldon will look *after* he finishes shaving, turn the mirror around, put it back on the dotted line, and look into it from the right.

18

Find the Duck

The rabbit is chasing a duck. Where's the duck?

19

Three-letter Word

In your mind, move the three arrows so they point to three letters that, taken left to right, spell a familiar word.

20

Four More Eye Twiddlers

The Bulgy Balloon

Is the balloon a perfect circle?

Two Men on a Cliff

One of these men must be suspended in mid-air, but which one?

Find the Center

Which dot is the true center of this circle? Take a guess before you make any measurements.

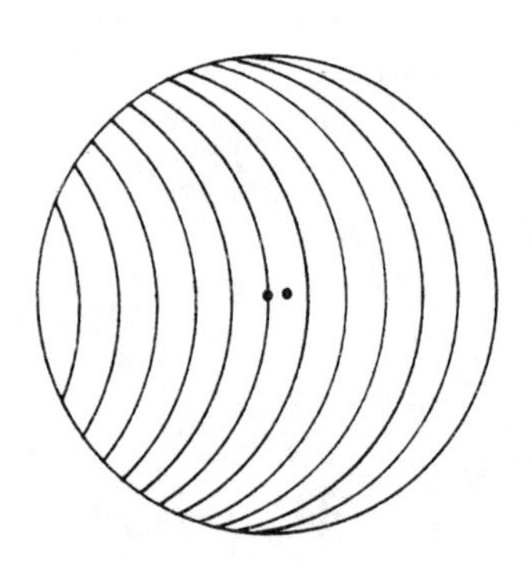

Spooky Spots

Do you see little gray spots at the intersections of the white lines between the black squares? When you try to look directly at a spot, it vanishes!

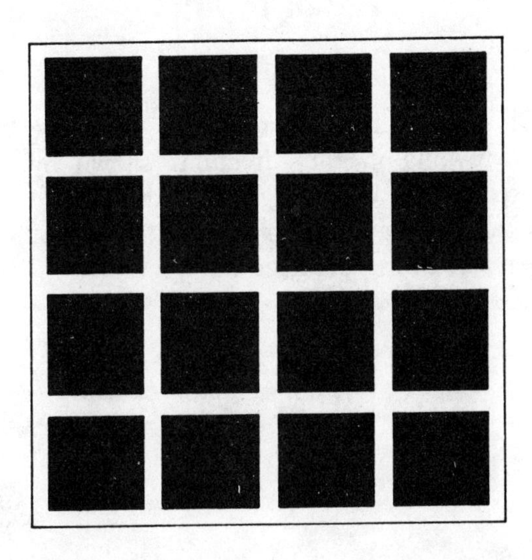

21

A Pair of Ants

On each of these twisted paper strips are two ants crawling in the directions shown by the arrows. If they keep on going in the same direction, and never cross an edge of the strip, will either pair ever meet head-on?

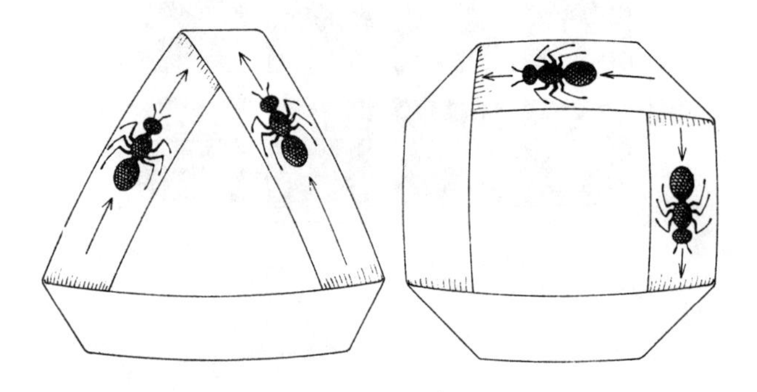

22

What Do You Say Next?

These are catches like the ones in the previous puzzle called "What Do You Do Next?," except that now it's what you *say* next that is the joke. Before you look at the answers, see if you can anticipate the joke and guess what you're supposed to say.

1. YOU: Did you know I have ESP? Go across the room, write any word on a piece of paper, fold it, put it under your left foot, and I'll tell you what's on the paper.
FRIEND: Okay, it's done. What's on the paper?

2. YOU: T-O, T-O-O, and T-W-O are all pronounced the same way. How do you pronounce the second day of the week?
FRIEND: Tuesday.

3. YOU: Ice in water makes iced water. What does ice in ink make?
FRIEND: Iced ink.

4. YOU: Touch your head and say the two letters for the abbreviation for "mountain."

FRIEND: M.T.

5. YOU: What's the difference between a mailbox and a garbage can?

FRIEND: I don't know.

6. YOU: If you asked your mother to put a stamp on an envelope and she refused, what would you do?

FRIEND: I'd put it on myself.

7. YOU: There's nothing I love to eat more than updock.

FRIEND: What's updock?

8. YOU: Just imagine! Getting up in the middle of the night to go horseback riding!

FRIEND: Who did that?

9. YOU: Seven times 8 is 54. Will you give me a dollar if I'm wrong?

FRIEND: Sure.

10. YOU: Suppose you're standing in line to buy an airplane ticket. The man in front of you is flying to London. The lady behind you is flying to Paris. Where are *you* going?

FRIEND: I give up.

11. YOU: It's running down my back!
FRIEND: What is?

12. YOU: Can you say "I one a bug, I two a bug, I three a bug," and so on, up through number eight?

FRIEND: I one a bug, I two a bug, I three a bug, I four a bug, I five a bug, I six a bug, I seven a bug, I eight a bug.

23

Mysterious Hieroglyphics

What do these strange symbols mean?

24

Tricky Mysteries

Murder at the Ski Resort

A Chicago lawyer and his wife went to Switzerland for a vacation. While they were skiing in the Alps, the wife skidded over a precipice and was killed. Back in Chicago an airline clerk read about the accident and immediately phoned the police. The lawyer was arrested and tried for murder.

The clerk did not know the lawyer or his wife. Nothing he'd heard or seen made him suspect foul play until he read about the accident in the paper. Why did he call the police?

Funny Business at the Fountain

At a hotel in Las Vegas, a lady rushed out of the manager's office to get a long drink at the water fountain in the lobby. A few minutes later she came out for another drink. This time she was followed by a man.

There was a mirror behind the fountain. When the lady raised her head, she saw that the man behind her had a knife in his upraised fist. She screamed.

The man lowered his knife, and then both of them began to laugh. What on earth is going on?

Accident on the Thruway

Mr. Jones was driving along the thruway with his son in the front seat. The road was icy. When Mr. Jones rounded a curve, his car skidded and rammed into a telephone pole. Mr. Jones was unhurt, but the boy broke several ribs.

An ambulance took the boy to the nearest hospital. He was wheeled into the emergency operating room. The surgeon took one look at the patient and said, "I can't operate on this boy. He's my son!"

How could this be?

25

Bee on the Nose

A bee has just landed on the girl's nose. Put a mirror's edge on the dotted line and look into it from the left to see how the girl looked a moment later.

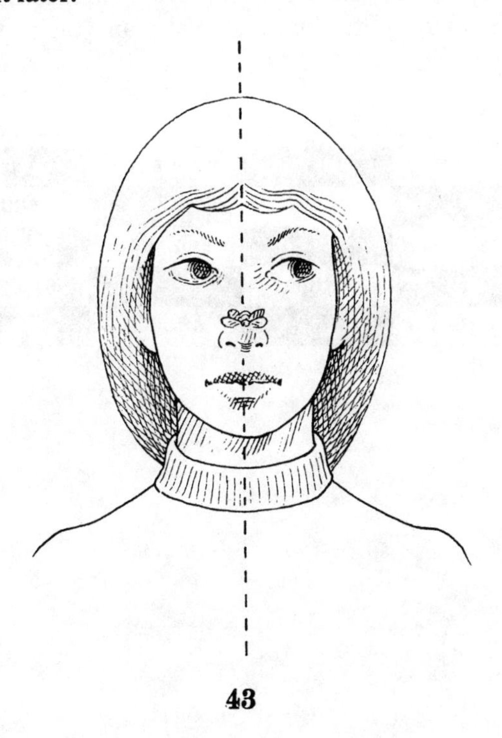

26

The Black and the White

Half the words below are black; half are white. Place the edge of a mirror on the dotted line above the words and look at the reflection of the words in the mirror. You'll see that the mirror reverses only the black words! Can you explain why?

CHOICE	PURPLE	COOKBOOK	WATER
WAR	DIED	TIGER	ECHO
ICE BOX	SQUARE	BOO HOO	TABLE
TURTLE	HIDE	LARGE	DECIDED
OBOE	ROSE	CHOKED	PIG

27

Lewis Carroll's Gift

On December 11, 1868, Lewis Carroll wrote the following letter to Dolly Argles, one of his child-friends:

My Dear Dolly,

. . . I'm going to send your Papa a little present this Christmas, which I daresay you may like to look at. It consists of some thin slices of dried vegetables that somebody has found a way of preparing so that they don't fall to pieces easily. They are marked in a sort of pattern with some chemical stuff or other, and fastened between sheets of pasteboard to preserve them. I believe the *sort* of thing isn't a new invention, but the markings of these are quite new. I inserted them myself. . . .

No more at present from

Your loving friend,

C. L. Dodgson

Dodgson was Lewis Carroll's real name. Can you guess what sort of gift Carroll sent to Dolly's father?

28

What's the Only Word?

There is one word, and one word only, that can be put inside each of the ten blank rectangles to give ten different meanings to the sentence.

The word must be added only once at a time. For example, if you put the word in the first rectangle, the meaning of the sentence changes. Move the same word to the second rectangle. Again the sentence makes sense, but now it means something still different. The same thing happens with the word in any of the other blanks.

What's the word?

[ ] TOM [ ] HELPED [ ] MARY'S [ ] DAUGHTER [ ] CLEAN [ ] MARY'S [ ] PARROT'S [ ] CAGE [ ] YESTERDAY [ ].

29

The Vanishing Moustache

Mr. Fulves is standing on the shore of a lake, watching the bathers. In that last picture it looks as if Mr. Fulves' moustache has mysteriously disappeared. Can you explain what happened?

30

The Three Kittens

The owner of this pet shop is trying to persuade two mothers and two daughters, who entered the store together, to buy the three kittens you see playing behind the glass.

He made the sale. Each customer left the store with her own pet kitten. None of them shared a kitten. There were no other kittens in the shop except the three you see here.

How can three kittens be owned by two mothers and two daughters so that each has her own individual pet?

31

Rescuing a Robin

A baby robin, trying to learn to fly, has tumbled into a hole in a large cement block that is part of the foundation of a building. The rectangular hole is just big enough to take a hand and arm, but it is more than three feet deep. No one can reach down far enough to pick up the baby bird. The construction workers don't know what to do. They're afraid to use long sticks for fear of hurting the bird.

Susan, who lives nearby, has just thought of a clever idea. When the men tried it, it worked beautifully. The robin was rescued unharmed. One of the men put a ladder against the tree and returned the baby robin to its nest.

What idea did Susan think of?

SAND

32

The Secret Code

To translate this coded message, read it in a mirror with its edge on the dotted line.

33

The Riders of Inviz

In an earlier puzzle we saw what happened in Inviz when all the musical instruments became invisible. The pictures on pp. 54-55 show what happened on the day that no one could see the object on which each was riding. What's invisible in each picture?

34

Rodney's Darts

Rodney has been throwing darts for hours to see if he can score exactly 100. As you see, he's just tossed four darts for a score of $17 + 17 + 17 + 24 = 75$ points, but now there is no way he can toss more darts to raise his score to 100.

Can you find a pattern of darts that will add up to exactly 100? You may use as many darts as you like. (Hint: Try first for a score of 50.)

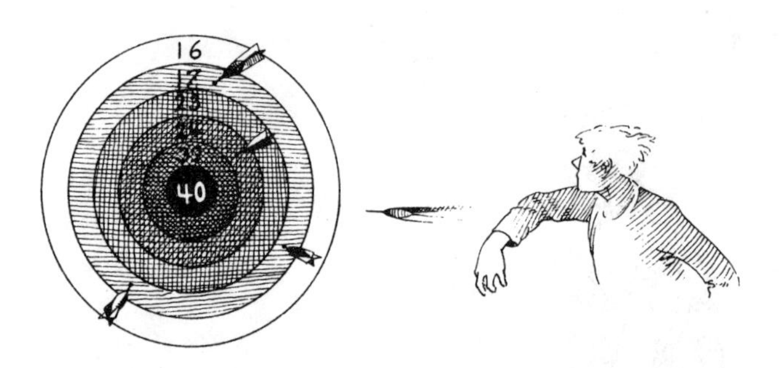

35

More What Do You Say Next?

Here are another dozen catches like the last twelve. The dialogue is between you and a friend. The puzzle is to guess what you say next.

1. YOU: What's the difference between a cow and a pocket calculator?

FRIEND: I give up.

YOU: Your old-age pension.

FRIEND: I don't get it.

2. YOU: What's the sum of 4Q and 6Q?

FRIEND: Ten Q.

3. YOU: It was Hank who taught me that last one.

FRIEND: Hank who?

4. YOU: Would you like to buy a potphur?

FRIEND: What's a potphur?

5. YOU: Did you know that Walter F. Doonton invented doontanite?

FRIEND: What's doontanite?

6. YOU: They're after you! They're after you!
FRIEND: Who?

7. YOU: There's a new joke about a girl who stood on one leg while she sang "The Star Spangled Banner." Have you heard it?
FRIEND: No.

8. YOU: I'm saving up my money to buy a henway.
FRIEND: What's a henway?

9. YOU: I'll bet you a dollar you won't answer "stick of gum" to three questions.
FRIEND: Okay. (Each of you put a dollar on the table.)
YOU: My first question is: "What's your name?"
FRIEND: Stick of gum.
YOU: What's the day of the week?
FRIEND: Stick of gum.

10. YOU: Say the word "two" twice, and then say the word "twain."
FRIEND: Two, two, twain.

11. YOU: Do you eat ice cream with your right or your left hand?

FRIEND: My right hand.

12. YOU: Did you know that people who are smart and beautiful are usually a trifle hard of hearing?

FRIEND: No, I didn't know that.

36

Noah's Ark

The picture shows the animals, two by two, lined up to enter Noah's Ark just before it began to rain. One pair of animals is out of place. Which pair is it?

37

Ms. Feemster's Message

These four pieces of aluminum sculpture, mounted on a wooden base, are the work of an abstract artist named Elinor Louise Feemster. She is trying to say something. But what?

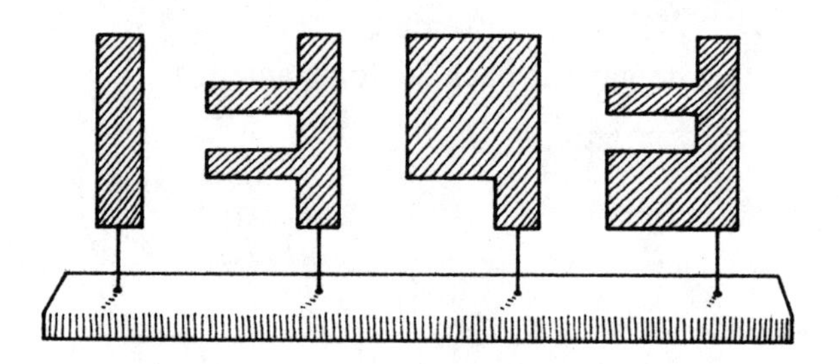

38

The Box Without a Lid

Lewis Carroll is the author of the following puzzle poem:

> John gave his brother James a box:
> About it there were many locks.
>
> James woke and said it gave him pain;
> So he gave it back to John again.
>
> The box was not with lid supplied,
> Yet caused two lids to open wide:
>
> And all these locks had never a key—
> What kind of box, then, could it be?

39

The Fish That U-turned

This charming toothpick puzzle comes from Japan. Arrange eight toothpicks to make the fish shown below. Add a dime for the fish's eye.

The problem: move the dime and just three toothpicks to make the fish swim in the opposite direction.

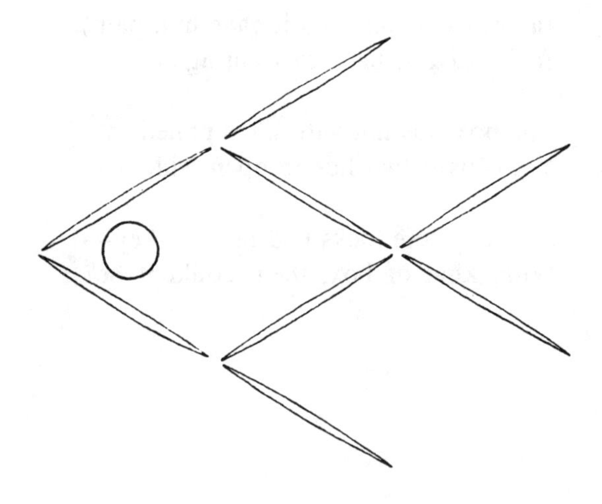

40

A New Angle on "What's the Difference?"

The artist has drawn the same scene as it would appear at the same moment from two different angles. But he's changed the scene slightly so there are seven spots where things are not the same. (See pp. 66-67). Can you find them?

41

Love and What?

Robert Clark is an artist who changed his last name to Indiana to honor his native state. In 1964 he became famous for his design of the word LOVE. The small picture shows how the word looked on a U.S. postage stamp in 1972.

The large picture is one of Mr. Indiana's aluminum sculptures. It, too, is telling us something in a single word. The puzzle is to guess what the word is.

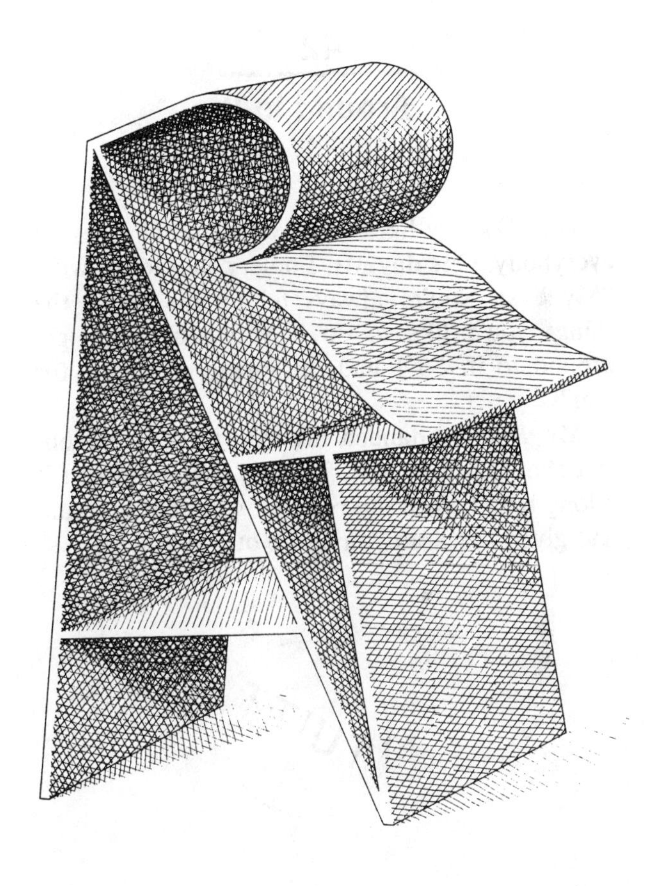

42

Around to It

Tom Foolery likes to play whimsical jokes on everybody. One day Mr. Foolery said to his wife, "My dear, you're always telling me about the things you'd like to do if you could only get around to it. So I had this made especially for you."

Mr. Foolery handed his wife the circular object shown below. Mrs. Foolery thought about it a long time before she realized what her husband had given her. Can you figure out what it is?

43

Where Do They Live?

Each person lives in a state that can be spelled by rearranging the letters of that person's name. For example, Roy Kewn lives in New York. Where do the others live?

Roy Kewn
Nora I. Charlton
Colin A. Fair
Dora K. Hatton
Earl Wade
Hilda D. Rosen
A. K. Barnes
J. R. Sweeney

44

The Flatz Beer Goof

Mr. Flatz, owner of the Flatz Beer Company in Milwaukee, decided to paint a slogan on the back of all his beer trucks. The slogan appeared on every truck exactly as shown in the picture.

After a few weeks everybody in Milwaukee was laughing about it. When Mr. Flatz discovered the reason, he had the slogan removed immediately from all the trucks.

Can you figure out why the slogan was so embarrassing to Mr. Flatz?

FLATZ
IS OUR
FINEST BEER
THERE IS NO
BEER SO GOOD
L. KUB
NY NY

45

Walking in the Rain

Mr. Brown left his apartment on Main Street, walked east on Main to the barber shop, got a haircut, then walked back home. The eight panels are not in proper sequence. Study them carefully and see if you can put them in the right time sequence. To start you off, the first panel is marked A.

DRUGS
DRUGS
DRUGS
A

46

Who Is It?

This marvelous picture was made with a technique developed by scientists at Bell Telephone Laboratories. It is a portrait of a very famous American.

To see who it is, have someone hold the book so you can look at the picture from a distance of 15 or 20 feet.

Answers

1
Ten Ridiculous Riddles

1. It quacks up.
2. A pink carnation.
3. Hold its nose.
4. "Men, get in the boat!"
5. He's a sloppy pizza eater.
6. A Mexican watermelon.
7. At the ghost office.
8. The.
9. A 5,000-pound sandwich that sticks to the roof of your mouth.
10. A mouse going on a long trip.

2
Fat Bats and Other Funny Beasts

Square bear
Half Giraffe
Jolly polly
Kickin' chicken

3
"The Whistler"

The Smiths have the painting upside down. If you turn the page around, you'll see the whistler. The label is the painting's number.

5
Tricky Questions

Speedy Retirement
It was daytime.

Peculiar Word
ISLAND.

Mystery Ball
Toss the ball in the air.

Shrewd Barbers

They make ten times as much money.

6

WHAT DO YOU DO NEXT?

1. Stick out your tongue and touch your nose with a finger.

2. If your victim says "yes," say, "Then I'll do it again." Do it again and repeat your question: "Did you like that?" If the victim says "no," say, "Then I'll take it back." Do it again, but this time brush your hand over his face from chin to forehead.

3. Write the words "red" and "blue."

4. Pick up his right hand as if it were a telephone, hold its fingers to your ear, and say "Hello."

5. Push your index finger through the cup's handle and give the plate a shove.

6. Bend the match in the center, like a V, before you drop it.

7. Put the newspaper sheet under a door. Have the two people stand on it, but on opposite sides of the door.

8. Put the empty pill bottle on the floor in the center of a room. Go out of the room and crawl back on your hands and knees. You are crawling "in to" the bottle!

9. When your friend extends her palm, smear it with lipstick.

10. Hold a burning match under a glass of water.

11. When the friend gives up, start humming.

12. Just keep quiet and don't do anything.

7
THE MUSICIANS OF INVIZ

The invisible instruments are:
Piano
Flute
Cello
Trombone
Harp
Violin
Cymbals
Drums
Guitar
Clarinet
Conga
Trumpet
(See pp. 84-85).

8

To Be or Not to Be

Baby
Back
Background
Bacon
Badge
Bag
Baggage
Baker
Bakery
Balcony
Bale
Ball
Balustrade
Band
Bananas
Bandage
Banjo
Banner
Barber
Bark
Barn
Barrel
Basin
Basket
Bat
Bellows
Belt
Bench
Bicycle
Billows
Biplane
Bay
Bay window
Beach
Beacon
Beak
Beard
Beast
Bed
Bedroom
Beef
Beets
Beggar
Belfry
Bell
Bird
Blacksmith
Blade
Blanket
Blinders
Blinds
Blindman
Board
Boat
Body
Bonnet

Book
Boot
Bough
Bow
Bowl
Bowsprit
Box
Boy
Braces
Bracket
Braid
Branch
Bread
Breeches
Bricks
Bridle
Brig
Bridge
Brook
Broom
Brush
Bubbles
Bucket
Buckle
Buggy
Buildings
Buns
Bundle
Buoy
Bureau
Bush
Butcher
Buttons

9
What's the Difference?

Circles are drawn around the six spots where the two pictures are not alike.

10
Easy as ABC

1. The Christmas message is "Noel" (no L).

2. The letters on the line are formed with straight line segments. Those below the line have curves.

3. Cab, cede, deed, fed, fee, feed, moon, noon, poop, pompon, rust, rusts, rut, ruts, strut, struts, tut, tuts, tut-tut, tut-tuts, tutu (a short skirt that ballerinas wear), tutus. Perhaps you can find some more.

11

Four Eye Twiddlers

The Two Spirals

The single rope is on the left.

The Curious Cube

The spots show the two right angles on the cube.

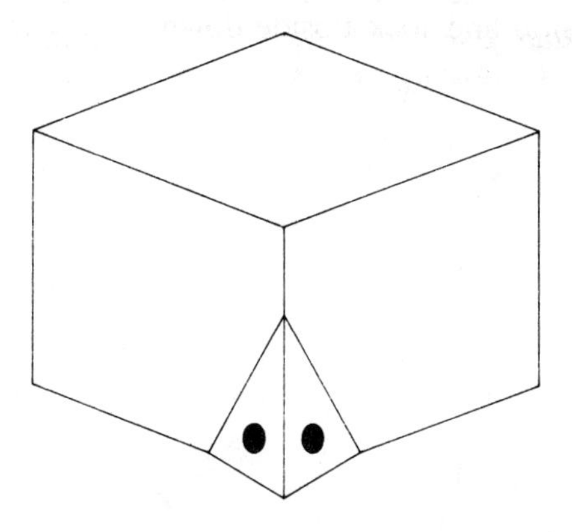

12
Dividing the Cake

Dad proposed that Henry and his sister flip a coin to see who cuts the cake. Then the person who does *not* cut the cake gets to choose the piece he or she wants.

13
Find the Mistakes

Zipper on door
Upside-down door
Doorknob and lock upside down
Double-barreled shotgun with three barrels
Mouse with squirrel's tail
Shirt on backward
Tie on backward
Hat on backward
Plants growing upside down
Men drinking out of bottom of bottle
Cat with five legs
Fisherman hooks bird
Sails backward
Barn floating on water
Plane flying upside down
Fish in birdcage
Snake with two heads
Upside-down glass of water suspended under the table

Extra finger on a hand
Upside-down umbrella cover
Smoke blowing one way, flag another
No holes in watering-can spout
Upside-down cup handle
Shoes not matching
Title of book on back cover
No trigger on gun
Upside-down drops from watering can

14

Three Match Puzzles

1. Move one match as shown to make a tiny square hole at the center.

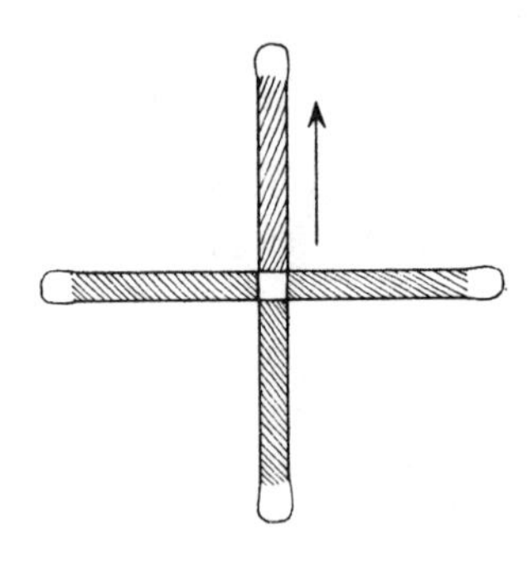

2. Move one match to change the position of the equal sign.

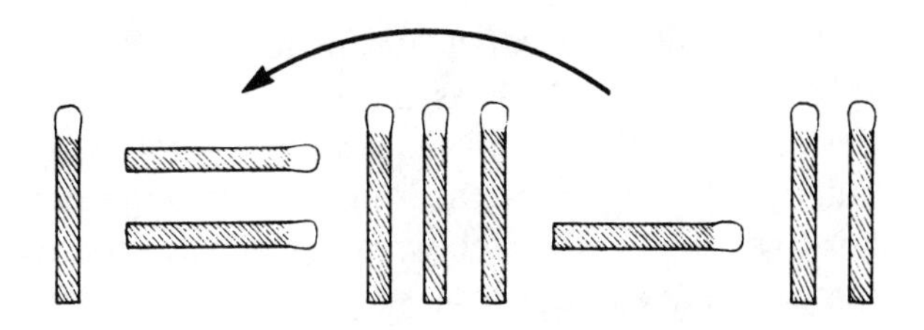

3. To make the equation correct, walk around the table and look at the matches from the other side!

15

Help Sherlock Catch the Criminals

The man is standing at the bottom of the picture and is the third person from the left. The woman also is standing in the foreground and is the fifth person from the left.

16

THE TWO WATERING CANS

The small can will hold the most water. Its spigot extends as high as the top of the can, so the can can be filled all the way to the top without overflowing from the spigot. The large can has a short spigot, which will overflow as soon as the can is filled above the spigot.

18

FIND THE DUCK

Give the page a quarter-turn to the left.

19

THREE-LETTER WORD

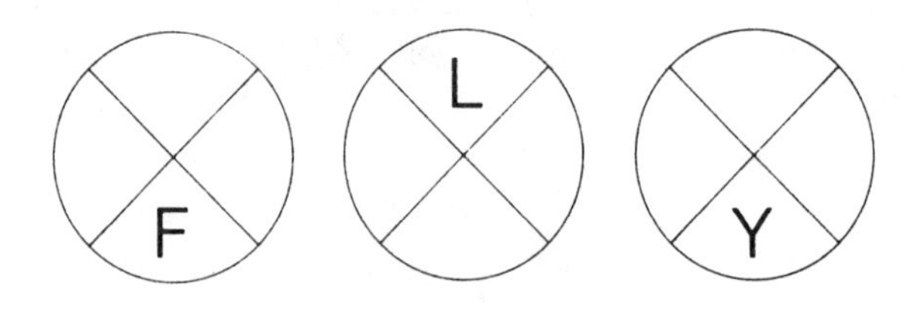

20
Four More Eye Twiddlers

The Bulgy Balloon

You can prove the balloon is a perfect circle by putting a quarter on top of it.

Two Men on a Cliff

The two cliffs form an "impossible figure" that can't exist in the real world. Cover either side with your hand and everything looks okay, but the entire picture doesn't make sense. Either man can be regarded as on a cliff, but then the other man will be in midair.

Find the Center

The dot on the left is the circle's center.

21
A Pair of Ants

The strip on the left is called a Moebius strip. It has only one side and one edge. In a moment the ants will meet head-on at the bottom of the strip. The other strip is a two-sided surface. Since the ants are on opposite sides, it is impossible for them to meet.

Take two strips of paper and paste the ends together so that they look exactly like the two pictures. Now try to cut each loop in half by cutting along the middle and all the way around. You'll be surprised by what happens!

22
WHAT DO YOU SAY NEXT?

1. "*You're* on the paper."
2. "That's funny. I always thought it was pronounced *Monday*."
3. "You should bathe more often."
4. "You said it."
5. "I'll never give you a letter to mail."
6. "Wouldn't it do more good to put the stamp on the envelope?"
7. "Who do you think you are, Bugs Bunny?"
8. "Paul Revere."
9. "Okay, I'm wrong. Where's the dollar?"
10. "Well, if you don't know where you're going, what are you standing in line for?"
11. "My spine."
12. "Really? How did it taste?"

23
Mysterious Hieroglyphics

If you imagine a straight line down the middle of each of the symbols, you'll realize that on the left of each line is a capital letter, and on the right is its mirror reflection. The letters are S, M, T, W, T, F, S—the initials of the days of the week.

24
Tricky Mysteries

Murder at the Ski Resort

The clerk had sold the lawyer a round-trip ticket to Switzerland, and a one-way ticket for his wife.

Funny Business at the Fountain

The lady had the hiccups. Her boss was trying to stop them by frightening her.

Accident on the Thruway

The surgeon was the boy's mother.

26

THE BLACK AND THE WHITE

The black words are unchanged by the mirror because they are spelled with letters that are unchanged when mirror-reflected and turned upside down. This is not true of the white letters.

27

LEWIS CARROLL'S GIFT

The Christmas present was one of Carroll's books.

28

WHAT'S THE ONLY WORD?

The only word is "only."

29

The Vanishing Moustache

Mr. Fulves doesn't have a moustache. What looks like the ends of his moustache in the first three panels are the ends of the canoe he is watching!

30

The Three Kittens

Who said there were *four* ladies? A woman came into the pet shop with her daughter and her granddaughter. That makes two mothers and two daughters!

31
Rescuing a Robin

Susan suggested that sand be poured very slowly, a tiny bit at a time, into the hole. The baby bird moved its feet to stay on top of the sand until the sand brought it up high enough in the hole to be reached.

33
THE RIDERS OF INVIZ

They are riding:
Horse
Bicycle
Canoe
Roller skates
Sled
Skis
Car
Motorcycle
Parachute
Balloon
Airplane
(See pp. 102-103).

34
Rodney's Darts

The only way to score 50 is with two darts on the 17 ring and one on the 16 ring. Do this twice —four darts on 17 and two on 16—and you will have a score of 100. If you're good at mathematics, you might enjoy proving there is no other way to score 100.

35
More What Do You Say Next?

1. "You will when you're 65."
2. "You're welcome."
3. "You're welcome."
4. "To cook in."
5. "I dunno. Let's check today's newspaper."
6. "V, W, X, Y, and Z."
7. "You should have. It's our national anthem."
8. "About seven pounds."
9. "Which would you rather have—the two dollars or a stick of gum?"
10. "Good. Now I'll teach you how to say 'airplane.' "
11. "Really? I use a spoon."
12. "Sorry, but I didn't quite hear you."

36
Noah's Ark

The animals are supposed to be in alphabetical order: alligators, bears, camels, ducks, elephants, frogs, giraffes, and so on. The frogs are in the wrong spot.

37
Ms. Feemster's Message

Place sheets of dark paper above and below Ms. Feemster's sculpture. You'll see that it spells her initials, E.L.F.

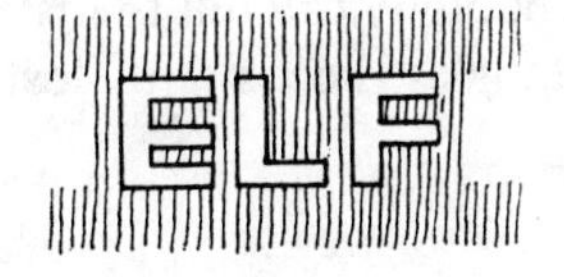

38
THE BOX WITHOUT A LID

Lewis Carroll answered his puzzle poem with another poem:

As curlyheaded Jemmy was sleeping in bed,
His brother John gave him a blow on the head;
James opened his eyelids, and spying his brother,
Doubled his fist, and gave him another.
This kind of box, then, is not so rare;
The lids are the eyelids, the locks are the hair,
And so every schoolboy can tell to his cost,
The key to the tangles is constantly lost.

39
THE FISH THAT U-TURNED

The dime and three toothpicks are moved as shown.

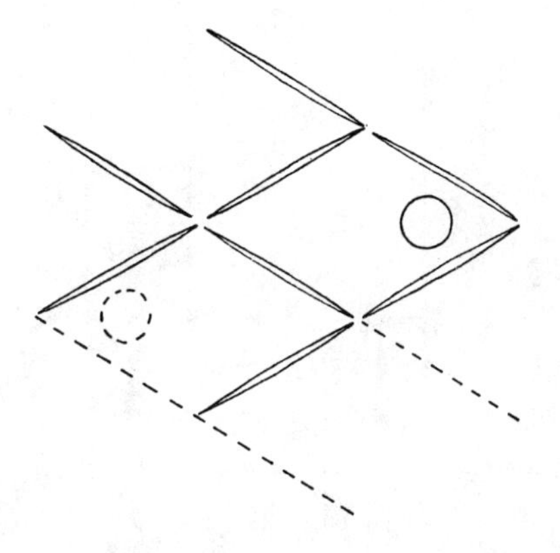

40
A NEW ANGLE ON "WHAT'S THE DIFFERENCE?"

The circles show the seven spots where things in the two scenes are not the same. (See pp. 108-109).

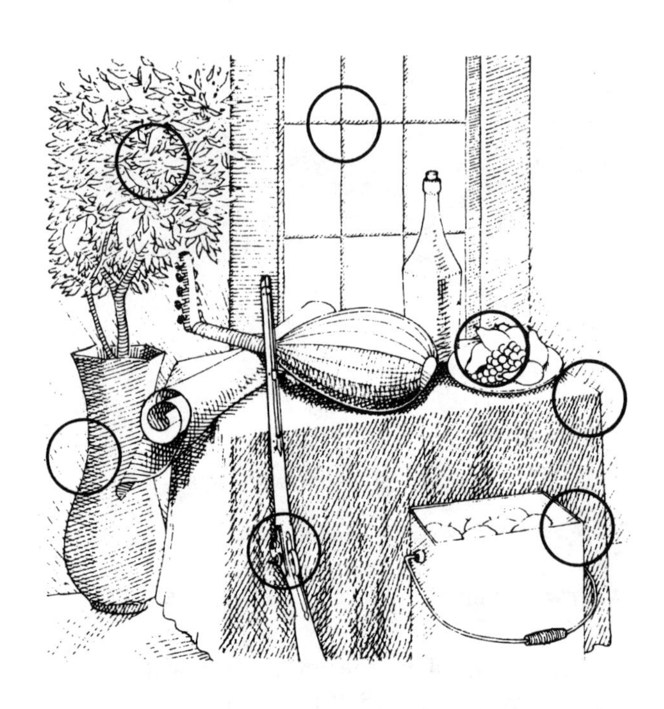

41
Love and What?

The sculpture is a huge monogram of the word ART.

42
Around to It

Tom Foolery gave his wife a "round tuitt." Go back and read what he said, and you'll understand. You can have some fun by making round "tuitts" out of cardboard. Hand them out to friends, telling them you heard they planned to do all sorts of things if they only got *a round tuitt*.

43

WHERE DO THEY LIVE?

Roy Kewn lives in New York; Nora I. Charlton in North Carolina; Colin A. Fair in California; Dora K. Hatton in North Dakota; Earl Wade in Delaware; Hilda D. Rosen in Rhode Island; A. K. Barnes in Nebraska; and J. R. Sweeney in New Jersey.

44

THE FLATZ BEER GOOF

When the doors on the back of the truck were opened, everybody could read a different slogan on the right side!

45

Walking in the Rain

A. Brown starts out without his umbrella. The sun is shining as he passes the drugstore.

B. The sun is still out when he passes the church.

C. It starts to rain.

D. Brown turns around and goes back for his umbrella.

E. He gets the umbrella, then starts out for the barber shop again. It is raining as he passes the drugstore.

F. It stops raining. The sun comes out and Brown closes his umbrella. He is passing the church.

G. On his way home, after getting his hair cut, it is still not raining as he passes the church.

H. When he reaches the drugstore it has started to rain again, so he opens his umbrella.

A
DRUGS
B
C
D
E
DRUGS
F
G
H
DRUGS

A CATALOG OF SELECTED

DOVER BOOKS

IN ALL FIELDS OF INTEREST

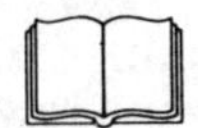

A CATALOG OF SELECTED DOVER BOOKS IN ALL FIELDS OF INTEREST

CATALOG OF DOVER BOOKS

AMERICAN CLIPPER SHIPS: 1833–1858, Octavius T. Howe & Frederick C. Matthews. Fully-illustrated, encyclopedic review of 352 clipper ships from the period of America's greatest maritime supremacy. Introduction. 109 halftones. 5 black-and-white line illustrations. Index. Total of 928pp. 5⅜ × 8½.
25115-2, 25116-0 Pa., Two-vol. set $17.90

TOWARDS A NEW ARCHITECTURE, Le Corbusier. Pioneering manifesto by great architect, near legendary founder of "International School." Technical and aesthetic theories, views on industry, economics, relation of form to function, "mass-production spirit," much more. Profusely illustrated. Unabridged translation of 13th French edition. Introduction by Frederick Etchells. 320pp. 6⅛ × 9¼. (Available in U.S. only) 25023-7 Pa. $8.95

THE BOOK OF KELLS, edited by Blanche Cirker. Inexpensive collection of 32 full-color, full-page plates from the greatest illuminated manuscript of the Middle Ages, painstakingly reproduced from rare facsimile edition. Publisher's Note. Captions. 32pp. 9⅜ × 12¼. 24345-1 Pa. $4.95

BEST SCIENCE FICTION STORIES OF H. G. WELLS, H. G. Wells. Full novel *The Invisible Man*, plus 17 short stories: "The Crystal Egg," "Aepyornis Island," "The Strange Orchid," etc. 303pp. 5⅜ × 8½. (Available in U.S. only)
21531-8 Pa. $4.95

AMERICAN SAILING SHIPS: Their Plans and History, Charles G. Davis. Photos, construction details of schooners, frigates, clippers, other sailcraft of 18th to early 20th centuries—plus entertaining discourse on design, rigging, nautical lore, much more. 137 black-and-white illustrations. 240pp. 6⅛ × 9¼.
24658-2 Pa. $5.95

ENTERTAINING MATHEMATICAL PUZZLES, Martin Gardner. Selection of author's favorite conundrums involving arithmetic, money, speed, etc., with lively commentary. Complete solutions. 112pp. 5⅜ × 8½. 25211-6 Pa. $2.95

THE WILL TO BELIEVE, HUMAN IMMORTALITY, William James. Two books bound together. Effect of irrational on logical, and arguments for human immortality. 402pp. 5⅜ × 8½. 20291-7 Pa. $7.50

THE HAUNTED MONASTERY and THE CHINESE MAZE MURDERS, Robert Van Gulik. 2 full novels by Van Gulik continue adventures of Judge Dee and his companions. An evil Taoist monastery, seemingly supernatural events; overgrown topiary maze that hides strange crimes. Set in 7th-century China. 27 illustrations. 328pp. 5⅜ × 8½. 23502-5 Pa. $5.95

CELEBRATED CASES OF JUDGE DEE (DEE GOONG AN), translated by Robert Van Gulik. Authentic 18th-century Chinese detective novel; Dee and associates solve three interlocked cases. Led to Van Gulik's own stories with same characters. Extensive introduction. 9 illustrations. 237pp. 5⅜ × 8½.
23337-5 Pa. $4.95

Prices subject to change without notice.

Available at your book dealer or write for free catalog to Dept. GI, Dover Publications, Inc., 31 East 2nd St., Mineola, N.Y. 11501. Dover publishes more than 175 books each year on science, elementary and advanced mathematics, biology, music, art, literary history, social sciences and other areas.